FORTSCHRITTE DER PHYSIK

HERAUSGEGEBEN IM AUFTRAGE DER PHYSIKALISCHEN GESELLSCHAFT
DER DEUTSCHEN DEMOKRATISCHEN REPUBLIK
VON F. KASCHLUHN, A. LÖSCHE, R. RITSCHL UND R. ROMPE

28. BAND 1980

AKADEMIE · VERLAG · BERLIN

Inhalt des 28. Bandes

Heft 1

Heft 2

Heft 3

Heft 4

Heft 5

Heft 6

Heft 7

Heft 8/9

Heft 10

Heft 11

Heft 12

FORTSCHRITTE DER PHYSIK

HERAUSGEGEBEN IM AUFTRAGE DER PHYSIKALISCHEN GESELLSCHAFT
DER
DEUTSCHEN DEMOKRATISCHEN REPUBLIK
VON F. KASCHLUHN, A. LÖSCHE, R. RITSCHL UND R. ROMPE

HEFT 1 · 1980 · BAND 28

AKADEMIE-VERLAG · BERLIN

EVP 10,— M
31728

BEZUGSMÖGLICHKEITEN

Bestellungen sind zu richten

- **in der DDR** an das Zeitungsvertriebsamt, an eine Buchhandlung oder an den AKADEMIE-VERLAG, DDR - 108 Berlin, Leipziger Straße 3–4
- **im sozialistischen Ausland** an eine Buchhandlung für fremdsprachige Literatur oder an den zuständigen Postzeitungsvertrieb
- **in der BRD und Westberlin** an eine Buchhandlung oder an die Auslieferungsstelle KUNST UND WISSEN, Erich Bieber, 7 Stuttgart 1, Wilhelmstraße 4–6
- **in Österreich** an den Globus-Buchvertrieb, 1201 Wien, Höchstädtplatz 3
- **in den übrigen westeuropäischen Ländern** an eine Buchhandlung oder an die Auslieferungsstelle KUNST UND WISSEN, Erich Bieber GmbH, CH - 8008 Zürich/Schweiz, Dufourstraße 51
- **im übrigen Ausland** an den Internationalen Buch- und Zeitschriftenhandel; den Buchexport, Volkseigener Außenhandelsbetrieb der Deutschen Demokratischen Republik, DDR - 701 Leipzig, Postfach 160, oder

  an den AKADEMIE-VERLAG, DDR - 108 Berlin, Leipziger Straße 3–4

Zeitschrift „Fortschritte der Physik"

Herausgeber: Prof. Dr. Frank Kaschluhn, Prof. Dr. Artur Lösche, Prof. Dr. Rudolf Ritschl, Prof. Dr. Robert Rompe, im Auftrag der Physikalischen Gesellschaft der Deutschen Demokratischen Republik.
Verlag: Akademie-Verlag, DDR - 108 Berlin, Leipziger Straße 3–4; Fernruf: 22 36 221 und 22 36 229; Telex-Nr. 114420; Bank: Staatsbank der DDR, Berlin, Konto-Nr. 6836-26-20712.
Chefredakteur: Dr. Lutz Rothkirch.
Anschrift der Redaktion: Sektion Physik der Humboldt-Universität zu Berlin, DDR - 104 Berlin, Hessische Straße 2.
Veröffentlicht unter der Lizenznummer 1324 des Presseamtes beim Vorsitzenden des Ministerrates der Deutschen Demokratischen Republik.
Gesamtherstellung: VEB Druckhaus „Maxim Gorki", DDR - 74 Altenburg, Carl-von-Ossietzky-Straße 30/31.
Erscheinungsweise: Die Zeitschrift „Fortschritte der Physik" erscheint monatlich. Die 12 Hefte eines Jahres bilden einen Band.
Bezugspreis je Band 180,– M zuzüglich Versandspesen (Preis für die DDR: 120,– M). Preis je Heft 15,– M (Preis für die DDR: 10,– M)
Bestellnummer dieses Heftes: 1027/28/1.

AN (EDV) 57618

ISSN 0015 - 8208

Fortschritte der Physik 28, 1—33 (1980)

The Non-Linear Geometric Optics of the Localized Wave Fields

A. B. SHVARTSBURG

Institute of Terrestrial Magnetism, Ionosphere and Radio Waves Propagation, Academy of Sciences USSR[1])

Abstract

The new approach to the self-action theory of intensive localized pulses, based on the hydrodynamical analogy in the non-linear geometrical optics, is proposed. The complex of phenomena of amplitude-phase non-stationary evolution of the intensive localized electromagnetic wave pulses in the dispersive medium is analysed in the framework of such approach. The wide classes of exact analytical solutions of the non-linear self-action equations, connected with such pulses, are constructed. The simple form of these solutions, represented with the well-known eigen-functions of the Laplace equation in special variables, permits to divide the pulse non-linear deformation qualitatively different effects. These solutions predict the large-scale pulse self-stratification and the origin of the quick intensity increase area during the non-linear evolution of the initially smooth distribution of the wave. The characteristic points of such evolution are represented by the singularities in the exact solutions of the non-linear geometrical optics. All results, describing the dynamics of the non-linear amplitude-phase re-building of the pulse, are represented in the simple algebraic form.

Contents

[1]) IZMIRAN, Akademgorodok p/O 142092, Moscow Region, USSR

Introduction

The interest for the wave processes, the wave amplitudes being large, is increasing during the last years in different fields of physics. Recently such processes attracted attention in the gas and fluid mechanics only. Afterwards, in connection with the appearance of the power radiation sources in the radio and optical ranges, the complex of problems of non-linear optics and non-linear electrodynamics began to extend. The investigations of processes of non-linear waves form the important trend in radio-physics, acoustics, and essentially in plasma physics. The considerable variety of plasma medium parameters causes the wide class if phenomena, associated with the non-linear dependence of the electromagnetic fields in plasma, even in case of relatively weak fields. Such investigations are of considerable theoretical and practical interest in connection with the problems of wave electronics, pecularities of the physical phenomena in powerful electromagnetic fields, and energy transfer. Probably, the beauty and the uncommonness of the non-linear equations and the phenomena, predicted by these equations, play the important role.

The analysis of wave processes shows a considerable analogy to the behaviour of definite wave modes in the non-linear medium, regardless of the different physical nature of these waves. The wide range of analysed problems stimulates the creation of a common approach and mathematical technique in order to describe such phenomena. It is interesting to note, that forty years ago the similar situation arised in connection with the analyses of non-linear oscillations of concentrated systems in radio technique, mechanics, and astronomy. Such situation led to the elaboration of the common theory of non-linear oscillations. In the last few years the progress of analysis of dynamics of intensive waves indicates the creation of the wave theory in non-linear distributive systems as a new independent field of physics.

The analogy in the tendencies of non-linear evolution of different wave fields permits to analyse these tendencies in the frame-work of the specific wave processes group without loss of generality. In this connection it is worth to concentrate attention to the problems of non-linear electrodynamics. Just in this trend the considerable progress, associated with laser and plasma experiments, opened the perspectives of theoretical prediction and controlled development of non-linear wave processes. The evolution of the intensive electromagnetic field in the non-linear medium leads to the origin and development of the wide class of phenomena, the dynamics of which is connected with the characteristics of various wave fields: intensity, polarization, spectrum, and space-time structure. The role of individual radiation parameters in the dispersive properties of the medium may be described for every concrete wave process with help of a special function — the non-linear permeability of the medium. The experimental and theoretical description of this function represents a very complicated problem [*1*].

In the non-linear electrodynamics perturbations of the dispersive properties of the medium in the wave field change the propagation conditions of the wave itself. The calculation of the influence of perturbations on the wave field leads to the equations of non-linear evolution of the electromagnetic field. The calculation of non-linear perturbations in the real and in the imaginary refractive index parts correspond to different processes. The perturbations of the imaginary part describe the effects of the non-linear absorption [*2*], the generation of harmonics [*3*], the cross-modulation of waves [*4*], and the parametrical excitement of oscillations [*5*]. The probabilities of such processes are usually calculated in the simple case, when the wave front may be approximated as a plane and the distribution of the intensity along the front is assumed to be constant. The above mentioned effects depend on the intensity of the local field only and are connected neither with the non-uniformity of the distribution of the intensity, nor with the phase structure of the field.

Unlike this, the perturbations of the real part of the refractive index characterize the non-linear corrections to the phase and group wave velocities. The role of such perturbations is essential in the evolution of the localized wave fields with non-uniform space time structure. This development of perturbations is possible to stimulate re-building of the radiation intensity and orientation in the non-linear medium. Just such effects characterize the space- localized propagation of wave beams and lead to the qualitatively new phenomenon — the non-linear radiation self-modulation phenomena, includes the self-constriction of the pulses, the broadening of the spectrum, and the self-stratification of the initially smooth wave distributions [*7*]. Such effects are possible not only in the fields of electromagnetic waves. Analogous phenomena are found in wave fields of quite different physical nature. It is worth while to show the qualitatively close effects, connected with the self-action of acoustic beams [*8*], the mutual focussing of light and sound [*9*], the self-constriction of distributed ion-acoustic waves and helicons [*10*], the deformation of a wave packet in a non-linear wave-guide [*11*], the self-localisation of perturbations in superfluid helium [*12*], in the crystal lattice [*13*], and in the solar wind plasma [*14*].

It is necessary to emphasize, that the above-mentioned division of the phenomena of non-linear electrodynamics into two groups, connected with the real and imaginary parts of the refractive index perturbations correspondingly, is to a marked degree conventional because both types of perturbations are connected in many phenomena. Thus the cumulation of wave energy as result of the perturbation of the real part of the refractive index causes the amplification of the wave transformation phenomena, which in their turn, are described by the perturbation of the imaginary part of the refractive index. However, such division seams to be useful in this rewiew, since the non-linear phenomena, connected with the imaginary part of the refractive index, will not discussed here. Recently these phenomena were described in the books, devoted to the quasi-particles interaction in the continuous medium [*15*], and, in particular, to the picture of statistical wave processes in a turbulent plasma [*2*].

Unlike this, the dynamical self-consistent picture of re-building of amplitude and phase profiles of localized pulses, connected with perturbations of the real part of the refractive index, is considerably less elaborated. The interest for this picture is stimulated by problems of controlled evolution of pulses of intensive waves. The rate and the dynamics of such evolution displays the essential dependence on the initial profiles of the pulses. Such dependence intensifies the interest for the exact analytical solutions of the equations of wave evolution, which permit to predict the amplitude-phase profile of the pulse at some moment of evolution, if the initial distribution of the wave is known. Therefore, the development of the localized non-linear waves non-stationary evolution theory seams to be very actual. The successes and the difficulties of this theory are the problems, the present review is devoted to.

The method of non-linear geometrical optics in application to the above-mentioned problems, stands out through this review. It is well known, that the linear approximation of geometrical optics is valid in the theory of optical instruments, wave propagation problems in geophysics, WKB method in quantum mechanics. Unlike this, the non-linear generalization of geometrical optics, connected with the self-consistent phase and intensity wave field deformation picture, is much more complicated. The wide classes of exact analytical solutions of the non-linear electromagnetic wave field self-action equations will be constructed here in the framework of the method of non-linear geometrical optics. On the basis of these solutions the tendencies of formation of the localized wave fields with the definite space-time structure will be analysed. The complex influence of the physical properties of the medium and of the radiation parameters on the non-linear wave field dynamics will be illustrated in the framework of this exact analytical method.

1. The non-linear re-building of the localized wave field

The non-stationary re-building of localized wave group in the homogeneous transparent non-linear medium will be discussed in this part. The non-linearity is supposed to be connected with the dependence of dielectric permeability on the electric wave field E only; the evolution of intensive wave in the medium with the non-linear magnetic permeability is discussed in [*38*]. In the isotropic medium the non-linearity may be described by a small additive term to the dielectric permeability "linear" value ε_0,

$$\varepsilon = \varepsilon_0 + \Delta\varepsilon.$$

The non-linear term $\Delta\varepsilon$, in the wide classes of the solid and liquid dielectric substances, [*16, 18, 19*], is proportional to the wave intensity W:

$$\Delta\varepsilon = \beta \cdot W.$$

The constant β is characterizing the substance given. The different types of electrodynamical non-linearities may be represented by the same formula [*17*]. It is worth while to note, that the dynamics of the pulse is very sensitive to the concrete type of the non-linear function $\Delta\varepsilon(W)$. The above-mentioned "quadratic" model represents the simplest case, formally connected with the first term in the expansion of the function $\Delta\varepsilon(W)$ in a power series in W. Other types of functions $\Delta\varepsilon(W)$, distorting the "quadratic" model effects, will be discussed below too.

1.1. The space-time field structure in the non-linear medium

This part is devoted to the equations, which describe the self-consistent re-building of the wave in the one-dimensional non-stationary problem. Let us restrict ourselves to the case of linear polarized quasi-monochromatic transverse waves.
Such propagation of waves along the x-direction is governed with the non-linear wave equation: for the complex field amplitude $E = E(x, t)$

$$\frac{\partial^2 E}{\partial x^2} - \frac{1}{c^2} \frac{\partial^2 (\varepsilon_0 + \Delta\varepsilon) \cdot E}{\partial t^2} = 0. \tag{1.1}$$

We shall suppose in what follows, that the non-linear term in (1.1) is small ($|\Delta\varepsilon| \ll \varepsilon_0$). Such assumption permits to emanate from the amplitude E the "quickly" and "slowly" varying parts:

$$E = \frac{1}{2} \left[A(x, t)\, e^{i(kx - \omega t)} + \text{c.c.}\right]. \tag{1.2}$$

The "quickly" varying part is described by the exponential factor in (1.2); the vector k and the frequency ω are connected in (1.2) by the linear dispersive equation

$$k^2 = \frac{\omega^2}{c^2} \varepsilon_0. \tag{1.3}$$

The slow accumulation of the non-linear distortions is described by the amplitude $A(x, t)$. If the non-linear evolution path is considerable longer than the wave-length $\lambda = 2\pi k^{-1}$, the amplitude is governed by the "shortened" self-action equation:

$$i\frac{\partial A}{\partial x} - \frac{1}{2v_0^2}\frac{\partial v_0}{\partial \omega}\cdot\frac{\partial^2 A}{\partial \xi^2} + \frac{\omega\cdot\Delta\varepsilon}{2c\cdot\sqrt{\varepsilon_0}}\cdot A = 0. \tag{1.4}$$

Here $\xi = t - x\cdot v_0^{-1}$; v_0 ist the group velocity of the wave, calculated from the eq. (1.3). The dispersion of the wave connected with the derivative $\partial v_0/\partial\omega$ is calculated with help of eq. (1.3) too. The dependence of group velocity on the intensity of the wave is neglected. The example of such dependence account is discussed in [*20*].
It is useful to transform eq. (1.4) to dimensionless quantities. Denoting the maximum value of the initial amplitude's modul by A_0, we shall analyse the dimensional complex amplitude $a = A\cdot A_0^{-1}$. Let us introduce the new variables θ and q:

$$\theta = \frac{x}{v_0 T}; \qquad q = \frac{t - x\cdot v_0^{-1}}{T}.$$

Here T is the characteristical duration of pulse at the beginning of the evolution. Finally, we can write the dimensionless form of eq. (1.4):

$$\begin{gathered} i\frac{\partial a}{\partial \theta} + p\frac{\partial^2 a}{\partial q^2} + f(a)\cdot a = 0 \\ P = -\frac{1}{2v_0 T}\cdot\frac{\partial v_0}{\partial\omega}; \qquad f = \frac{v_0}{c}\cdot\frac{\omega T\cdot\Delta\varepsilon}{2\sqrt{\varepsilon_0}}. \end{gathered} \tag{1.5}$$

If the intensity of the wave is small $(f(a)\to 0)$, this equation turns into the linear parabolic equation, which is well known in the diffraction theory [*4*]. If the dispersion is weak $(p\to 0)$, this equation describes the phase distortion of the pulse, connected with the non-linear phase-velocity perturbation in the dispersionless medium. Unlike these extreme cases eq. (1.5) describes the evolution of the field, associated with both abovementioned factors — the dispersion and the non-linearity. It is supposed in eq. (1.5), that the non-linear group velocity perturbation cannot stimulate the considerable pulse deformation during the evolution processes in question.
The solutions of eq. (1.5) describe the evolution of the non-stationary space-time wave field, the initial phase and amplitude being known. Moreover, this equation characterizes the stationary re-building of the cylindrical wave which propagates along θ-direction in the non-linear medium (θ is the longitudinal co-ordinate, q is the transverse one, $p = 1$). However, we face the considerable mathematical difficulties, when we attempt to construct the solutions of this equation in the common case. Therefore, the role of results, which show the general tendencies of non-linear evolution of localized, is very important.
In order to reveal such tendencies, different methods were developed: it is worth while to mark the qualitative [*21*], the numerical [*22*], and the asymptotic [*23*] methods. The asymptotic stratification of the rectangular pulse, (the Fraunhofer diffraction in the nonlinear medium) was discussed in [*23*]. The essential general result, connected with the sign of product $p\cdot f$ in eq. (1.5) was founded by Lighthill [*24*]: according to this result, the solutions of eq. (1.5) are stable against the modulation type instabilities, in the case $p\cdot f < 0$: in the case $p\cdot f > 0$ one has the modulation instability. The equation (1.5) with the concrete functions p and f is frequently applied in the non-linear theory of the waves of different physical nature [*10*]. The essential progress of the analyse of this

equation is associated with the article of V. E. ZAKHAROV and A. B. SHABAD [25], devoted to reducing this equation to the inverse scattering problem. For the first time this method was used in [46] to the Korteweg-de Vries equation problems [10]. The algorhythm of solution of eq. (1.5), proposed in [25], reduces this equation to the linear integral equations set. However, the new difficulties, connected with the above-mentioned set solution, arise in the framework of such approach.

Thus, the absence of an useful analytical technique impeds the complete description of wave evolution in the non-linear dispersion medium. The pace of the evolution is determined by the competition between the processes of non-linear self-action and the dispersive spreading out. It is sufficient for evolution of intensive fields that the non-linear distortions of the field structure under the action of the inhomogeneity of the amplitude-phase distribution accumulate more rapidly than the dispersive spreading out of the initial field distribution occurs. Moreover, in a number of problems the size of region occupied by non-linear medium is limited so that the amplitude and the phase field distributions at the entrance to the medium play an important role in the field-self-action process inside the region. It makes sense to consider the simplified evolution problems, which are considered with the application of the geometrical optics.

1.2. The large-scale picture of the evolution of the non-linear wave field

The equations of the non-linear geometrical optics approximation may be derived from the non-linear Schrödinger equation (1.5). Let us examine the case when the time of establishment of the signal $\tau_0 = \sqrt{\pi|\partial^2 s/\partial\omega^2|}$ (s is the wave phase) is small in comparison with the duration of the pulse. Separating the real and the imaginary parts in the complex amplitude a in eq. (1.5), $\big(a = b \cdot \exp(is)\big)$, we find the coupling equations for the intensity $|b|^2$ and phase s of the wave.

Let us introduce the new variable τ and the new functions:

$$\tau = \theta\sqrt{2|p \cdot f_0|}; \qquad W = |b|^2; \qquad u = \sqrt{2\left|\frac{P}{f_0}\right|}\,\frac{\partial s}{\partial q}; \qquad F = \frac{|f|}{|f_0|}\,\delta_1 \cdot \delta_2;$$
$$p = |p| \cdot \delta_1; \qquad f = |f| \cdot \delta_2; \qquad \delta_{1,2}(y) = \left.\begin{matrix} +1 \\ -1 \end{matrix}\right\} \begin{matrix} y > 0 \\ y < 0 \end{matrix}. \tag{1.6}$$

Here f_0 is the maximum of the function $f(a)$ in eq. (1.5); the signs δ_1 and δ_2 correspond to the different types of the dispersion and the non-linear permeability. The eq. (1.5) transforms to the set of equations:

$$\frac{\partial W}{\partial \tau} + \frac{\partial (Wu)}{\partial q} = 0; \tag{1.7}$$

$$\frac{\partial u}{\partial \tau} + u\,\frac{\partial u}{\partial q} - \frac{\partial F}{\partial W} \cdot \frac{\partial W}{\partial q} - \frac{|p|}{|f_0|} \cdot \frac{1}{\sqrt{w}}\,\frac{\partial^2 \sqrt{W}}{\partial q^2} = 0. \tag{1.8}$$

If the wave's distribution is smooth enough,

$$q^2 \gg \left|\frac{p}{f_0}\right|, \tag{1.9}$$

the role of the last term in eq. (1.8), connected with the small-scale diffraction effects contribution to the amplitude evolution, may be negligible. Such disregard, which is characteristical in the non-linear geometrical optics, permits to limit oneself in the eq. (1.8) to the first derivatives only:

$$\frac{\partial u}{\partial \tau} + u \frac{\partial u}{\partial q} + \frac{\partial F}{\partial W} \cdot \frac{\partial W}{\partial q} = 0. \tag{1.10}$$

Eqs. (1.7) and (1.10) form the master system in the non-linear geometrical optics. Eq. (1.7) represents the transfer equation, and the equality (1.10) represents the non-linear eikonal equation. Two functions $W(\tau, q)$ and $u(\tau, q)$ describe the intensity and the phase structure of the field. Such approach is useful at the "geometrical optics" stage of the field's evolution. Such period of evolution depends essentially upon the initial amplitude-phase distribution of the pulse. The corresponding initial envelopes represent the boundary conditions to the set (1.7)—(1.10):

$$W|_{\tau=0} = W_0(q); \qquad u|_{\tau=0} = u_0(q). \tag{1.11}$$

The direct construction of exact solutions of the set (1.7)—(1.10) in the form of evident functions $W(\tau, q)$ and $u(\tau, q)$ at the arbitrary boundary conditions (1.11) case is unknown. Below the wide classes of the non-linear geometrical optics exact analytical solutions, satisfying the conditions (1.11) are constructed.

2. The exact solutions of the equations of non-linear geometrical optics

This paragraph is devoted to the new method of exact analytical solutions of non-stationary non-linear electrodynamical problems. This problem is connected with the system (1.7), (1.10). These equations are analogous to the hydrodynamical Eulerian equations which describe the arbitrary motion of the one dimensional compressible gas with the specific heats ratio $\gamma = 2$. In this analogy the wave intensity W and the phase gradient are similar to the hydrodynamical density and speed of stream; the non-linear permeability $F(W)$ corresponds to the equation of state in the gas dynamics. The significant difference between the set (1.7), (1.10) and the Eulerian equation is connected with the change of the "pressure" sign $\delta_1\delta_2 = \pm 1$ in eq. (1.10): the hydrodynamical pressure aspires to expand the stream; unlike this, the non-linear self-action aspires to constrict the distribution of the wave in the case $\delta_1\delta_2 = \pm 1$. The self-action sign change leads to the equation type change; the set (1.7), (1.10) is an elliptic one, when $\delta_1\delta_2 = +1$. In the case $\delta_1\delta_2 = -1$ this system, similar to the hydrodynamical equations, has the hyperbolic type. In spite of this, the common equation's likeness permits to utilize the hydrodynamical methods in the non-linear geometrical optics.

2.1. The hydrodynamical analogy

It is important, that the set of equations (1.7), (1.10) of non-linear geometrical optics admits to a wide class of exact analytical solutions. For the construction of such solutions it is convenient to transform this set to a linear one, regarding W and u as independent variables, and $\tau(W, u)$ $q(W, u)$ as functions:

$$\delta_1\delta_2\alpha \frac{\partial \tau}{\partial u} + u \frac{\partial \tau}{\partial W} - \frac{\partial q}{\partial W} = 0; \qquad \alpha = \frac{\partial F(W)}{\partial W}; \tag{2.1}$$

$$\frac{\partial q}{\partial u} - u \frac{\partial \tau}{\partial u} + W \frac{\partial \tau}{\partial W} = 0. \tag{2.2}$$

Unlike the initial system (1.7), (1.10) the set of equations (2.1), (2.2) is linear. Here the functions $\tau(W, u)$ and $q(W, u)$ are supposed to be single-valued. The equation (2.1) plays the role of eikonal equation, and eq. (2.2) plays the role of transfer equation in (W, u)-space.
The set of two first-order differential equations obtained on this way can be reduced to one second-order differential equation which expresses the quantities q and τ in terms of some function $\psi(W, u)$ through relations such as the Legendre transformations. In the approximation of geometrical optics this function contains all informations, connected with the evolution of the non-linear field. This can be accomplished in two different ways by choosing the function ψ to satisfy exactly the eikonal equation or the transfer equation in (W, u)-space. Of course, the eigen-functions ψ are different in each case. Depending on the actual form of the non-linear response of the medium and the initial shapes of the amplitude and frequency envelopes one or another method is the convenient one to use.

Transformation I satisfies exactly the transfer equation in (W, u) space. The functions $q(W, u)$ and $\tau(W, u)$ are introduced by the formulae:

$$\tau = \frac{\partial \psi_1}{\partial u}; \qquad q = -\psi_1 - W\frac{\partial \psi_1}{\partial W} + u\frac{\partial \psi_1}{\partial u}. \tag{2.3}$$

The equation for the function ψ_1, which follows from the eikonal equation, then has the form

$$\frac{\partial^2 \psi_1}{\partial b^2} + \frac{3}{b}\frac{\partial \psi_1}{\partial b} + \alpha\delta_1\delta_2\frac{\partial^2 \psi_1}{\partial v^2} = 0; \tag{2.4}$$

$$W = b^2; \qquad v = 2u. \tag{2.4a}$$

Transformation II, unlike transformation I, satisfies exactly the eikonal equation in (W, u)-space. The functions $\tau(W, u)$ and $q(W, u)$ are introduced by the formulae:

$$\tau = -\frac{1}{\alpha}\frac{\partial \psi_2}{\partial W}; \qquad q = -\frac{u}{\alpha}\frac{\partial \psi_2}{\partial W} - \delta_1\delta_2\frac{\partial \psi_2}{\partial u}. \tag{2.5}$$

The equation for the function ψ_2, which follows from the transfer equation, has the form:

$$\frac{\partial^2 \psi_2}{\partial b^2} + \frac{1}{b}\left[\frac{\alpha}{b}\cdot\frac{\partial}{\partial b}\left(\frac{b^2}{\alpha}\right) - 1\right]\frac{\partial \psi_2}{\partial b} + \alpha\delta_1\delta_2\frac{\partial^2 \psi_2}{\partial v^2} = 0. \tag{2.6}$$

The boundary conditions for this equation are connected with the structure of the pulse at the boundary of the non-linear medium ($\tau = 0$).
Functions ψ_1, ψ_2 being calculated, the field self-action description reduces to the formal differentiation procedure in accordance with formulae (2.3), (2.5). The "inverse" expressions $\tau(W, u)$ and $q(W, u)$ are constructed in this approach. The calculation of the "obvious" functions $W(\tau, q)$ and $u(\tau, q)$ is, as usual, impossible, but the treatment of the inverse functions $\tau(W, u)$ and $q(W, u)$ permits to analyse the evolution dynamics. First of all, such treatment discovers the characteristical moments of evolution, connected with the formation of singularities at the initial smooth wave's distribution.

2.2. Two tendencies at the evolution of the localized wave field

The gradients of the field may stimulate two tendencies in the non-linear development of pulse dynamics.

1. The slope of the pulse front increases or decreases monotonically. The increase of the slope leads to the appearance of a region with an appreciable gradient of the field. In this

region the derivatives $\partial W/\partial q$ and $\partial u/\partial q$ formally tend to infinity. Such behaviour is connected with the vertical tangent origin on the intensity profile $W(\tau, q)$. If such region of quick intensity growth arose within the region of wave localization, the condition of such singularity origin has the form

$$\left.\frac{\partial q}{\partial W}\right|_{\tau} = 0; \quad \left.\frac{\partial^2 q}{\partial W^2}\right|_{\tau} = 0. \tag{2.7}$$

This singularity is analogous to the spillover of a simple wave in hydrodynamics. If the above-mentioned region develops at the periphery of the waves distribution, the condition of such singularity origin is:

$$W = 0; \quad \left.\frac{\partial q}{\partial W}\right|_{W} = 0. \tag{2.8}$$

The formation of the region of quick growth of intensity may occur near the distribution maximum too. The profile of the pulse being symmetric $[W(q) = W(-q)]$, the condition of such singularity formation has the form:

$$q = 0; \quad \left.\frac{\partial q}{\partial W}\right|_{\tau} = 0. \tag{2.9}$$

Unlike the case (2.7), the singularity (2.9) has no hydrodynamical analogy.
So the non-linear evolution of the initially smooth wave profile stimulates the distortion of the field structure and the development of the envelopes quick changes regions. Such evolution stage is characterized formally by the origin of singularities in the self-action equation solutions. After such origin of singularities the equations of non-linear geometrical optics became invalid.

2. Unlike this monotonous regime, the tendency of wave distribution evolution is connected with the non-monotonous development. In this case the intensity profile slope decreases at any distribution part. Such processes development leads to the horizontal bend point formation at the intensity profile. The condition of such points origin has the form:

$$\left.\frac{\partial W}{\partial q}\right|_{\tau} = 0; \quad \left.\frac{\partial^2 W}{\partial q^2}\right|_{\tau} = 0, \tag{2.10}$$

after these bend points formation the self-stratification of the waves distribution evelops. The "geometrical optics" stage of such regime of non-monotonous evolution is described with help of eq. (1.7), (1.10) too.

3. The model of "quadratic" non-linear permeability

The number of non-linear wave self-action processes may be analysed in the simple "quadratic" approximation of non-linear permeability:

$$F(W) = W.$$

In this case the function $\alpha(W) = dF(W)/dW$ in eq. (2.4) and (2.6) is equal to 1. Namely, this model is utilized often to describe the evolution of light pulses in non-linear dielectric, Langmuir and ion-acoustic waves in plasma [*10*]. In this case the eq. (2.4) transfers to the five-dimensional axial-asymmetrical Laplace equation in the (b, v) space (v is the axis of symmetry):

$$\frac{\partial^2 \psi_1}{\partial b^2} + \frac{3}{b}\frac{\partial \psi_1}{\partial b} + \delta_1\delta_2 \frac{\partial^2 \psi_1}{\partial v^2} = 0. \tag{3.1}$$

Eq. (2.6) transfers to the well-known three-dimensional Laplace equation:

$$\frac{\partial^2 \psi_2}{\partial b^2} + \frac{1}{b}\frac{\partial \psi_2}{\partial b} + \delta_1\delta_2 \frac{\partial^2 \psi_2}{\partial v^2} = 0. \tag{3.2}$$

Eqs. (3.1) and (3.2) are very important in the non-linear geometrical optics. The success in construction of exact analytical solutions of this equation, satisfying the boundary conditions (2.3) and (2.5) correspondingly depends essentially on the choice of the coordinate system and the eigen-functions of these equations, associated with such system. The direct utilization of (b, v) co-ordinate system produces difficulties in the formulation of boundary conditions, in the case of localized wave pulses. It seems to be convenient in the case of two-dimensional non-linear electro-dynamical problems to transform the self-action equations to spheroidal co-ordinates. In this case the corresponding eigen-functions reflect the number of peculiarities of pulse's structure during the evolution adequate. Let us analyse the cases $\delta_1\delta_2 > 0$ and $\delta_1\delta_2 < 0$ in eq. (2.4), (2.6) separately.

3.1. The eigen-functions of the non-stationary problem

Let us introduce in the (b, v) space in the case $\delta_1\delta_2 = +1$ the spheroidal coordinates by the substitutions

$$b^2 = (1 + \varepsilon^2)(1 - \eta^2); \qquad v = \varepsilon \cdot \eta; \qquad \varepsilon \geqq 0; \qquad -1 \leqq \eta \leqq 1. \tag{3.3}$$

The solution $\psi_2(\varepsilon, \eta)$ of the Laplace equation can be written now in form of a series in terms of oblate spheroidal harmonics of first (P_n) and second (Q_n) kind:

$$\psi_2 = \sum_n \{Q_n(\eta)[A_nQ_n(i\varepsilon) + B_nP_n(i\varepsilon)] + P_n(\eta)[C_nQ_n(i\varepsilon) + D_nP_n(i\varepsilon)]\}. \tag{3.4}$$

The coefficients in this expansion can be determined from the boundary conditions; these conditions are specified on the $\varepsilon = 0$ surface. For the initially non-modulated pulse $(v|_{q=0} = 0)$ the boundary conditions have the form:

$$\left.\frac{\partial \psi_2}{\partial \eta}\right|_{\varepsilon=0} = 0; \qquad \left.\frac{\partial \psi_2}{\partial \varepsilon}\right|_{\varepsilon=0} = -2\eta \cdot q_0(W). \tag{3.5}$$

Here $q_0(W)$ is the inverse of the $W_0(q)$. Unlike the expansion (3.4) the solutions of the five-dimensional Laplace equation (3.1), connected with the transformation I, may be represented in a form:

$$\psi_1 = \sum_n \{T_n(\eta)[A_nT_n(i\varepsilon) + B_nR_n(i\varepsilon)] + R_n(\eta)[C_nT_n(i\varepsilon) + D_nR_n(i\varepsilon)]\}. \tag{3.6}$$

Here R_n and T_n are Jacobi polynomials and functions:

$$R_0(\eta) = 1; \qquad T_0(\eta) = \frac{1}{2}\left[\frac{\eta}{1-\eta^2} + L\right];$$

$$R_1(\eta) = \eta; \qquad T_1(\eta) = \frac{1}{2}\left[2 - \frac{\eta^2}{1-\eta^2} - 3R_1(\eta) \cdot L\right]; \tag{3.7}$$

$$R_2(\eta) = 1 - 5\eta^2; \qquad T_2(\eta) = \frac{1}{2}\left[15\eta - \frac{2\eta}{1-\eta^2} + 3R_2(\eta) \cdot L\right];$$

$$R_3(\eta) = 3\eta - 7\eta^3; \qquad T_3(\eta) = \frac{1}{16}\left[16 - 99\eta^2 + \frac{6\eta^4}{1-\eta^2} - 15R_3(\eta) \cdot L\right]; \tag{3.7}$$

$$L = \frac{1}{2} \ln \frac{1+\eta}{1-\eta}.$$

The boundary conditions in case (3.5) have the form:

$$\left.\frac{\partial \psi_1}{\partial \varepsilon}\right|_{\varepsilon=0} = 0; \qquad -\psi_1 + \left.\frac{1-\eta^2}{2\eta} \frac{\partial \psi_1}{\partial \eta}\right|_{\varepsilon=0} = q_0(\eta). \tag{3.8}$$

It is essential that the harmonics P_n and R_n and their derivatives are limited in the interval $(-1 \leqq \eta \leqq 1)$; the harmonics Q_n and T_n tend to infinity when $\eta^2 \to 1$ in the same interval. Soon we shall see, that the roles of these harmonics in the boundary conditions representation are of different quality. The representation of solutions of eq. (2.4), (2.6) with help of spheroidal harmonics is convenient, if the localized wave's field distribution has it's centre and decreases at it's periphery. On the contrary, the toroidal coordinate system and the toroidal harmonics in the (b, v) space are convenient, if the distribution reaches it's maximum intensity at it's periphery and the intensity decreases at the centre [*28*].

3.2. The hyperbolic equation of wave evolution

The type of self-action equation is connected with the dispersive and the non-linear properties of the medium. In the case $\delta_1\delta_2 = +1$ the above-mentioned equation belongs to the elliptic type; the case $\delta_1\delta_2 = -1$ corresponds to the hyperbolic type equation. It is useful to generalize the above-mentioned method of eigen-functions at the case $\delta_1\delta_2 = -1$ and to introduce the new variables (ε, η) at the (b, v) space by the formulae [*41*] $(\varepsilon \leqq 1)$:

$$b^2 = (1 - \varepsilon^2)(1 - \eta^2); \qquad v = -\varepsilon\eta. \tag{3.9}$$

The substitution

$$\varepsilon \to i\varepsilon, \qquad \frac{\pi}{2} - \operatorname{arc\,tan} \frac{1}{\varepsilon} \to \frac{1}{2} \ln \frac{1+\varepsilon}{1-\varepsilon}, \tag{3.10}$$

transfers the solutions (3.4), (3.6) to the corresponding solutions, associated with the hyperbolic case $\delta_1\delta_2 = -1$. In the framework of such approach the solution of the hyperbolic equation has the form of a sum:

$$\psi_2 = \sum_n \{Q_n(\eta)\,[A_nQ_n(\varepsilon) + B_nP_n(\varepsilon)] + P_n(\eta)\,[C_nQ_n(\varepsilon) + D_nP_n(\varepsilon)]\}. \tag{3.11}$$

The further analyse of solution (3.9) is quite analogous to the case $\delta_1\delta_2 = +1$. Solutions of eqs. (3.1) and (3.2) are represented here in form of the linear combinations of eigen-functions. From the other hand, the solution of eq. (3.1) in the case $\delta_1\delta_2 = +1$ was represented by M. Lighthill [*24*] in form of a contour integral in a complex plane (b, iv):

$$\psi = -\frac{i}{2} \int\limits_{b-i(v-v_0)}^{b+i(v-v_0)} \cdot\, db_0 \cdot q_0(b) \sqrt{\frac{b_0}{b}} \cdot F\left[\frac{1}{2}; \frac{1}{2}; 1; \frac{(v-v_0)^2 + (p-p_0)^2}{-4bb_0}\right]. \tag{3.12}$$

Here F is the hypergeometric function, $v_0 = v(q)|_{\tau=0}$; $q_0(b)$ is the inverse to $b_0(q) = b(q)|_{\tau=0}$. Such integral calculation faces the serious difficulties, connected with the complication of integration in the complex plane with the different initial distributions $q|_{\tau=0} = q_0(b)$. The integral (3.12) was calculated in [24] in one private case $W|_{\tau=0} = (1 + q^2)^{-2}$, associated with the hydrodynamical problem of wave group evolution on deep water. At the same time, the picture of field dynamics is very sensitive to the initial distribution $q_0(b)$ of envelope. The adequate description of field re-building demands the procedure of function ψ construction to be very adaptable. This procedure must reflect the influence of different details of the initial distribution easily.

Unlike the integral representation (3.12) the expansion approach of eigen-functions permits to unify the analytical investigation technique in application to a wide class of initial distributions $W_0(q)$. The standard analytical operations with eigen-functions of the Laplace equation permit to separate the initial envelope each harmonics deposit to the general evolution picture. Because of the set of harmonics, adequate to specific initial distribution type, it becomes possible to separate qualitatively the different effects of non-linear field evolution. Namely, such approach will be illustrated below.

Let us note in conclusion, that the solutions of non-linear electrodynamical discussed here, are not traditional in the theoretical physics. The evolution equations in the case $\delta_1\delta_2 = +1$ are of elliptical type, and in the case $\delta_1\delta_2 = -1$ their type is hyperbolic. As usually, the solutions of elliptic equations are represented with help of the expansion to eigen-functions, and the solutions of hyperbolic equations are constructed in the framework of the method of characteristics. However, the approach of non-linear geometrical optics, developed here, leads to the elliptic self-action equations with the boundary conditions of Cauchy type; such boundary conditions are associated with the hyperbolic problems. This situation is typical to the "uncorrect" problems of mathematical physics. In these problems the intensity and the modulation initial envelopes $W_0(q)$ and $v_0(q)$ are connected with their simultaneous existence condition: $W_0 = W_0(v_0)$.

In this case the arbitrary independent intensity and modulation distributions are inadmissible. If the mutual modulation is absent $\big(v_0(q) = 0\big)$, this condition can be dropped. Such "mixed" situation stimulates the unusual analytical technique: the elliptical equation (2.3) $\delta_1\delta_2 = +1$ is solved with help of characteristics (3.12) in the complex plane (b, iv). Simultaneously the hyperbolic equation (3.2) $(\delta_1\delta_2 = -1)$ is solved with help of eigen-functions (3.11). Namely, the "uncorrectness" of the problem stimulates the origin of singularities in the solutions of elliptical equations. It is essential, that such singularities connected with the origin of large field gradients, arise in the finite evolution time. After origin of such singularities the equations of non-linear geometrical optics lose their employment. However, the theory, which predicts the moment and the position of such singularity origin, connected with the initial distribution of field, is of significant theoretical and practical interest. The "uncorrect" problems attract attention in the different physical problems [*54*]. Because of the "uncorrect" problems approach the progress was achieved in the non-linear geometrical optics too.

4. The waves amplitude-frequency self-modulation non-stationary phenomena

The non-linear geometrical optics method of large-scale pulses self-action description is applied systematically in this paragraph to the concrete analyse of wave distributions. The pulse dynamics depend on the non-linear distortion and dispersive expansion processes competition. In particular, it is possible to find the special pulse envelopes, corresponding to such tendencies mutual compensation in every point of the field localization region. The existence of stationary solutions of the non-linear Schrödinger equation (1.5) is possible in this case. Within the framework of "quadratic" non-linear permeability model private solutions have the "soliton" form (W_1) in the case $\delta_1\delta_2 = +1$ [32] and the cavity form (W_2) in the case $\delta_1\delta_2 = -1$ [33]

$$W_1 = \coth^{-2}\left(\frac{q}{T_0}\right); \qquad W_2 = \tanh^2\left(\frac{q}{T_0}\right); \qquad T_0 = \left[2|p| \cdot \left|\frac{df(a)}{da^2}\right|^{-1}\right]^{1/2}. \tag{4.1}$$

The solitary waves (4.1) may propagate without distortion of the intensity envelopes. The stationary solution of the non-linear Schrödinger equation may have the form of persistent amplitude-modulated waves too [12]. Such waves are described with help of Jacobi elliptic functions.

The above-mentioned conditions of the dispersive and the non-linear effects mutual compensation distinguish the specific pulse profiles (4.1). In the common case such conditions are not fulfilled and the non-linear re-building of pulses develops. In the approximation of non-linear geometrical optics the re-building dynamics of pulses is illustrated often in the framework of two private examples. These examples were received directly from eqs. (1.7), (1.10), without utilization of the inverse functions $q = q(W, u)$, $\tau = \tau(W, u)$:

1) The parabolic distribution:

$$W|_{\tau=0} = \begin{matrix} 1 - q^2 \\ 0 \end{matrix} \left. \begin{matrix} \\ \\ \end{matrix} \right\} \begin{matrix} q^2 \leqq 1; \\ q^2 \geqq 1; \end{matrix} \qquad u(q)|_{\tau=0} = 0. \tag{4.2}$$

The profile (4.2) was investigated by V. I. Talanov [40]. His result predicts the collaps of the distribution into a "focus" near by the pulse centre ($q = 0$); the time of evolution τ_φ is

$$\tau_\varphi = -\frac{\pi}{4}. \tag{4.3}$$

Profile (4.2) is determined on the limited interval ($-1 \leqq q \leqq 1$) but leads to the non-limited intensity value at the quick field change region near by the "focus".

2) The other set (1.7), (1.10) private solution leads, unlike (4.2) to the finite intensity value at the region of quick change of field however, this solution describes the profile, distributed along the infinite interval [7]

$$W_0(q) = \coth^{-2} q. \tag{4.4}$$

The wide classes of exact analytical solutions determined on the limited interval of q values will be discussed below. These solutions lead to the values of finite intensity at the regions of quick change of field. The appearance of such regions corresponds to the origin of singularities in the solutions of non-linear geometrical optics equations.

The (ε, η) co-ordinate system (3.3) and the corresponding solutions (3.4), (3.6) will be utilized below. The well-known pulses (4.2) and (4.4) may be expressed easily with help of the simplest self-action equation (3.2) eigen-functions P_n and Q_n in the (ε, η) space. The initial parabolic distribution (4.2) is expressed by the formula:

$$q_0 = \sqrt{1 - W}; \qquad = \eta = P_1(\eta).$$

Analogously, the "soliton" initial distribution (4.4) has the form:

$$q_0 = \operatorname{arcoth}\frac{1}{\sqrt{W_0}} = \frac{1}{2}\ln\frac{1+\eta}{1-\eta} = Q_0(\eta);$$

These "inverse" dependences, expressed with help of only one Legendre function, may be represented in the direct forms (4.2) and (4.4) too. Probably, because of this simplified circumstance, the above-mentioned profiles are utilized long ago as standard solutions in the non-linear electrodynamics. Unfortunately, the more complicated initial distributions $q_0(W)$ associated with combinations of the linear eigen-functions P_n and Q_n, cannot be expressed, as usual, at the direct form $W_0(q)$. Namely, in these cases, such an approach, connected with the "inverse" equations (3.1) and (3.2), proves to be extremely helpful.

4.1. The self-constriction of the symmetric wave profile

Let us consider the pulse determined on the interval $-1 \leqq q \leqq 1$, which is characterized, in contrast to (4.2), by a smooth variation of intensity at the edges (Fig. 1):

$$W_0|_{q=1} = W_0|_{q=-1} = 0; \qquad \left.\frac{\partial W_0}{\partial q}\right|_{q=1} = \left.\frac{\partial W_0}{\partial q}\right|_{q=-1} = 0. \tag{4.5}$$

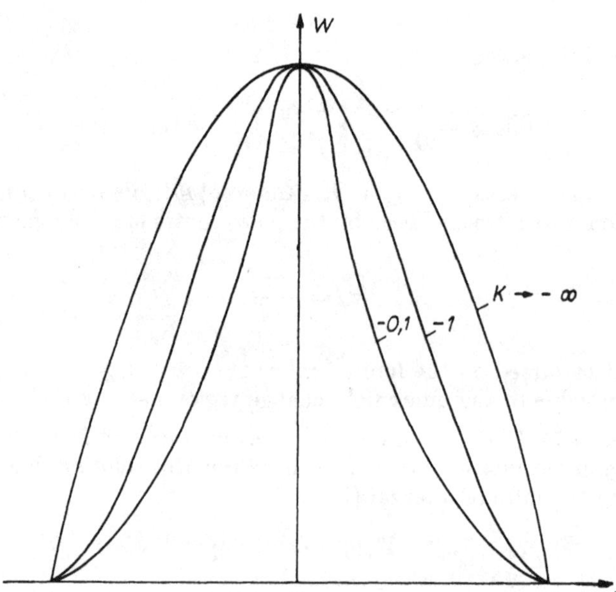

Fig. 1. Initial intensity distribution in the pulse for different values of parameter K (the symmetric profile)

Suppose, that the boundary condition, which specifies the intensity distribution $W_0(q)$ in q for $\tau = 0$ is given by

$$q = \eta + \frac{1-\eta^2}{2(K-1)} \ln \frac{1+\eta}{1-\eta}, \tag{4.6}$$

where K is a parameter and (3.3):

$$W_0 = W|_{\varepsilon=0} = 1 - \eta^2. \tag{4.7}$$

The function, given by (4.6), describes the one-parameter family of symmetric pulses $[W(q) = W(-q)]$, determined on the interval $(-1 \leqq q \leqq 1)$, where $q = \pm 1$ corresponds to $\eta = \pm 1$, respectively, and the intensity is maximal ($W_0 = 1$) near by the centre

($q = 0$; $\eta = 0$). To ensure that (4.6) is a single-valued function, it is required, that the equation $\partial q/\partial W = 0$ should have no roots for $-1 \leqq \eta \leqq 1$. The calculation yields, that this condition is fulfilled, if values of the parameter K are limited:

$$K < 0. \tag{4.8}$$

We shall assume henceforth, that (4.8) is satisfied. In this case by varying the value of K and using (4.7) we can investigate the self-action of a broad class of profiles. Substituting (4.6) into the boundary conditions (3.5) we obtain the following expressions for the coefficients in (3.4):

$$A_1 = \frac{4i}{5(K-1)}; \qquad A_3 = \frac{2}{3} A_1; \qquad C_0 = \frac{5}{6} A_1;$$
$$C_2 = -\frac{2i(K-2)}{3(K-1)}; \qquad D_2 = \frac{\pi}{3} \cdot \frac{K-2}{K-1}. \tag{4.9}$$

The remaining coefficients in (3.4) are zero. Hence, using (2.5), we obtain the following formulae for the required functions:

$$q = \frac{f_1(\varepsilon;\, \eta;\, K)}{K-1}; \qquad \tau = \frac{f_2(\varepsilon;\, \eta;\, K)}{K-1}; \tag{4.10}$$

$$f_1 = \frac{1+\varepsilon^2+\varepsilon^2\eta^2-\eta^2}{2} \ln\frac{1+\eta}{1-\eta}; \quad + \frac{\eta\varepsilon^4(1+3\eta^2)}{3(1+\varepsilon^2)\,(1-\eta^2)} + \frac{\eta(K-1)}{1+\varepsilon^2};$$
$$f_2 = 2\varepsilon\left[\frac{\eta}{2}\ln\frac{1+\eta}{1-\eta} - \frac{1}{1+\varepsilon^2} + \frac{\varepsilon^2(3\eta^2-2)}{(1+\varepsilon^2)\,(1-\eta^2)}\right]$$
$$+ \frac{K-2}{2}\left[\arctan\frac{1}{\varepsilon} - \frac{\pi}{2} - \frac{\varepsilon}{1+\varepsilon^2}\right]. \tag{4.11}$$

In these expressions the variables ε and η are related to W and v by (3.3).
The formulae (4.10)—(4.11) contain all informations about deformation of the nonlinear pulse (4.6) .On the base of these formulae we shall now consider the singularities, which appear during the evolution of pulse (4.6). Respectively to values of the parameter K, two types of singularities are possible:

a) The singularity in the maximum of intensity ($\eta = 0$):
The substitution of the solution (4.10), (4.11) to the condition (2.9) yields the critical variables ε_k value [*28*]:

$$\varepsilon_k{}^2 = \frac{3}{4}\left[-1 + \sqrt{1 - \frac{4K}{3}}\right]; \qquad (K < 0). \tag{4.12}$$

The substitution of (4.12) into the solutions (4.10), (4.11) illustrates the result of evolution, connected with the singularity origin near by the centre of the pulse (Fig. 2).

b) The perphery singularity corresponds to the quick intensity growth region origin at the periphery of the pulse. Such "spillover" point coordinates in the (ε, η) space are determined after the substitution [*28*] of the solution (4.10), (4.11) to the equations (2.7)

$$\eta_k{}^2 = 10^{-3}; \qquad \varepsilon_k{}^2 = \left(-\frac{2K}{3}\right)^{1/2}. \tag{4.13}$$

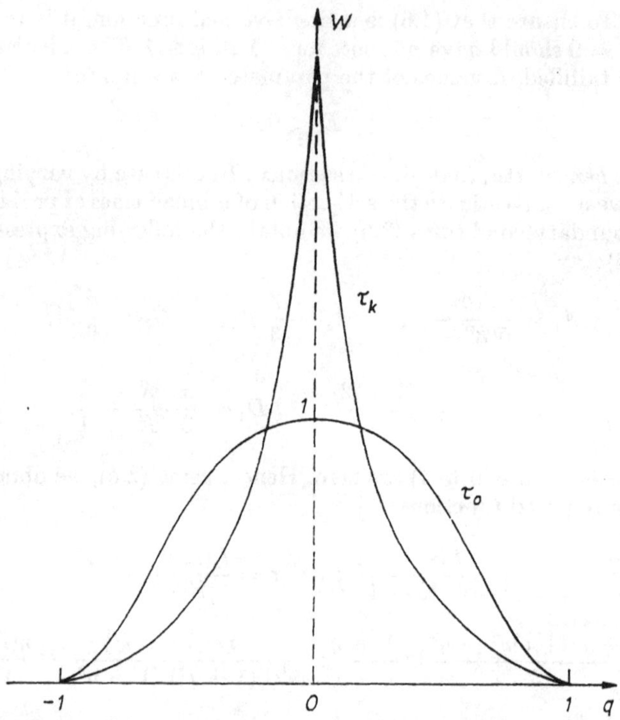

Fig. 2. The appearance of the quick intensity growth region near by the symmetric pulse maximum

The intensity in the singular point is calculated with help of (3.3):

$$W_k = (1 + \varepsilon_k^2)(1 - \eta_k^2). \tag{4.14}$$

In this case the evolution of the initial profile results in an distribution of intensity near the centre's maximum, which takes the form of a thin intensity spike with a smooth maximum (Fig. 3).

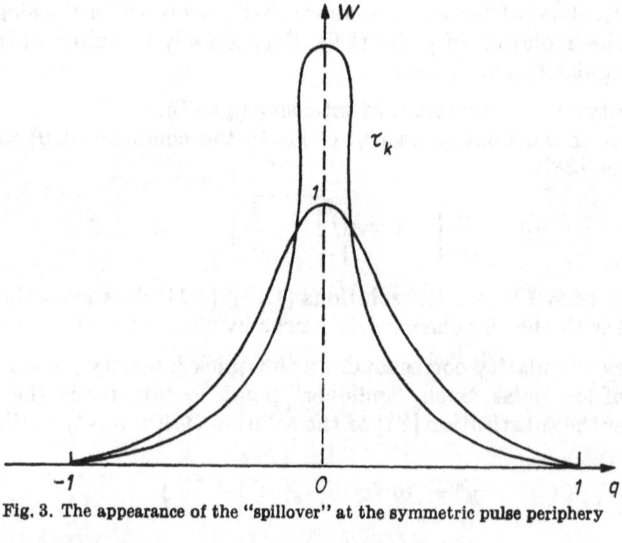

Fig. 3. The appearance of the "spillover" at the symmetric pulse periphery

The analyse of solutions (4.12) and (4.13) proves, that the non-linear dynamics of the family of pulses (4.6) depends the values of parameter K. Namely, the "spillover" is the first to appear if $K < -235$ whilst for $K > -235$, the centre singularity appears as first. This is readily understood if we recall, that for large values of $|K|$ the initial profile (4.6) is not very different from the parabolic profile (4.2), which has a tendency to self-constricting on the centre. For smaller values of $|K|(|K| < 235)$ the effect of the "edges" of the profile $W_0(q)$ leads to a qualitatively different result of pulses self-action — to an increase of the envelope steepness outside of the centre and the appearance of the "spillover" singularity.
It is worth while to note, that different pulses (4.6) may be represented as a result of "addition" of the parabolic profile with another distribution. Namely, the parameter determines the role of each distribution. When $K \to 0$, the deposit of parabola is small; the intensity is distributed on the whole in the thin region near by the maximum ($q = 0$); the gradients of the field are large, outside the centre of pulse. Unlike this, in the case $K \to -\infty$ the role of parabola becomes decisive.
It is essential, that the values of intensity in the singularity points, (4.14), connected with the limited variables and η_k values, are finite. The singularity is connected with the gradients of large pulses only. In the private case of unlimited growth of parameter K ($K \to -\infty$) the formula (4.12) describes the intensity growth type, corresponding to the parabolic pulse self-constriction, $W_k \to \infty$. So, the parabolic profile is exceptional; and any small distortion of such profile leads to the limitation of intensity at the singular point.
The above-mentioned solutions were analysed in the frame work of the eigen-functions of eq. (3.2). The utilization of eq. (3.1) eigen-functions is analogous. As an example we consider pulses envelopes at the boundary of the non-linear medium ($\tau = 0$), described by the function (Fig. 4).

$$q|_{\tau=0} = \frac{2\eta}{1-\eta^2} + \frac{3}{2} K\eta - \frac{20 - 30\eta^2 - K}{4} \ln \frac{1+\eta}{1-\eta}. \tag{4.15}$$

The relation (4.15) is single-valued if the parameter K values are limited by the condition $-14 < K < -10$. Substituting (4.15) into the boundary condition (3.8), we find the values of the coefficients in the representation (3.6)

$$\psi_1 = (2 - K)\, T_0(\eta) + (K - 10) \cdot R_1(\eta) \cdot T_1(i\theta) - T_2(\eta) \cdot R_2(i\theta). \tag{4.16}$$

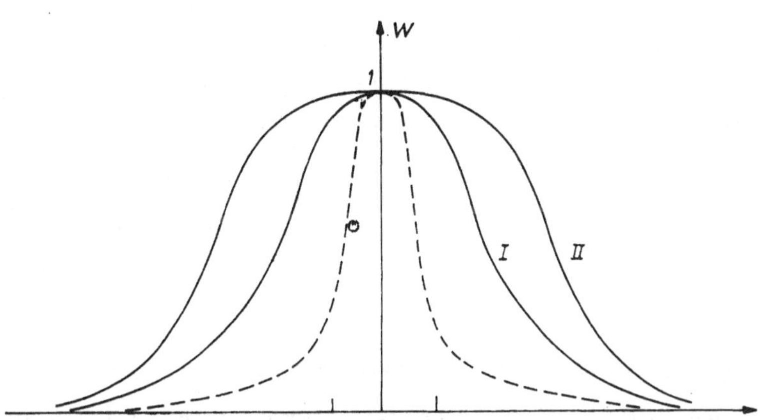

Fig. 4. Family of pulses, described with the help of five-dimensional Laplace equation (the dotted curve shows the pulse $W = \mathrm{ch}^{-2} q$; curve 1: $K = -11$; curve 2: $K = -13$)

Fig. 4. shows, that the pulse (4.15) decreases from the centre ($W_0 = 1$) to the periphery ($W_0 \to 0$) more slowly, than the "soliton" pulse $W_0 = \coth^{-2} q$, although their maximal intensities are equal. The singularities analyse of in the solution (4.16) is fulfilled with help of conditions (2.7), (2.9) too.
The above-mentioned solutions describe the convex profiles. The evolution of concave profiles in the case $\delta_1\delta_2 = +1$ was investigated in [28] with help of toroidal harmonics in the (b, v) space. The corresponding solution predicts an increase in the steepness at the edges of the profile and the appearance of singularity (2.8) (Fig. 5).

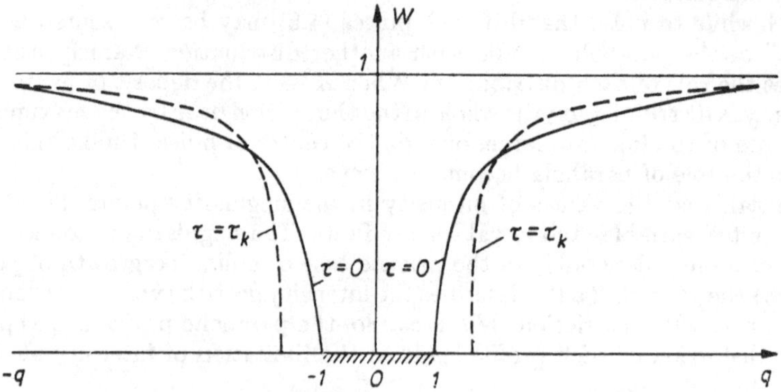

Fig. 5. The increase of steepness at the edges of the concave profile (the continuous curve shows the initial profile, the dotted curve illustrates the result of evolution)

4.2. The evolution of the asymmetric wave envelope

We now consider the self-action of an asymmetric pulse $W_0(q) \neq W_0(-q)$, which satisfies the conditions, given by (4.5). Let us consider the two-parameter family of initial profiles:

$$q|_{\varepsilon=0} = \frac{1-\eta^2}{2} \ln \frac{1+\eta}{1-\eta} = (K-1)\,\eta + M\eta^2. \tag{4.17}$$

Here K and M are parameters. In analogy to (4.10), (4.11) one can construct the expression $\tau(\varepsilon, \eta)$ and $q(\varepsilon, \eta)$ in the case (4.17):

$$q = f_1(\varepsilon; \eta; K) + M(\eta^2 - \varepsilon^2 - \varepsilon^2\eta^2); \qquad \tau = f_2(\varepsilon; \eta; K) - 2M\varepsilon\eta. \tag{4.18}$$

Here the functions f_1, and f_2 are determined by (4.10), (4.11). It is directly clear from the solution given by (4.18) that in the course of the evolution, the point corresponding to the maximum of intensity, is displaced away to the edge of the pulse, so that $\Delta q = -M\varepsilon^2$ (Fig. 6; $K = -9$; $M = 0{,}4$). The evolution of the pulse leads to an increase in the steepness and to "spillover" singularity formation. It is worth while to note, that the initial asymmetry of the envelope, expressed by the parameter M, is not limited to small values of M. So, the solution (4.18) is valid in the case of considerable maximum pulse displacement too, before formation of the singularity.

4.3. The non-linear pulse large-scale self-stratification

The monotonous self-constriction of pulse represents only one possible tendency of self-action. The other tendency of non-linear evolution is connected with a decrease of curvature at some region of the profile and to the forthcoming stratification of the pulse.

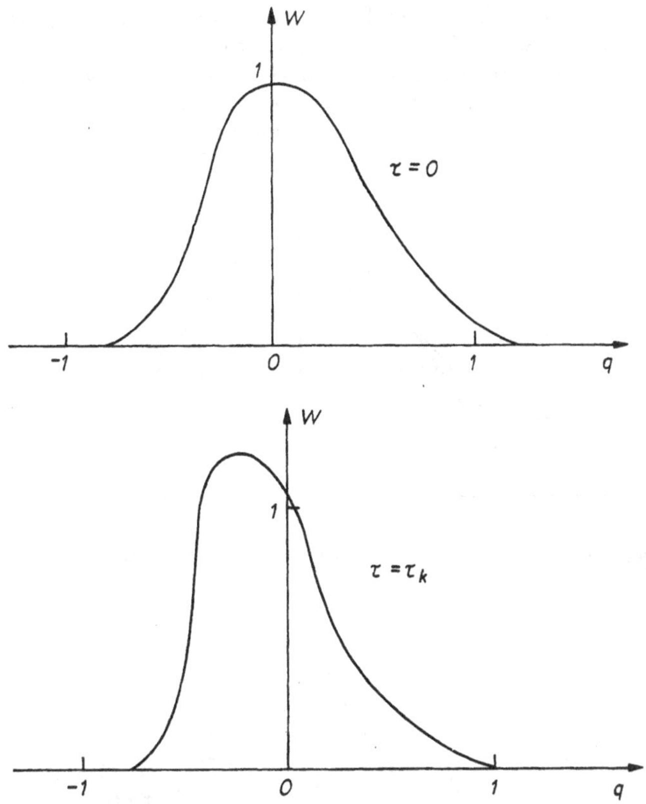

Fig. 6. The asymmetric profile evolution (fig. 6a shows the initial pulse, fig. 6b shows the asymmetric singularity formation)

The new maxima of intensity may arise both at the centre and at the periphery of wave distribution; the type of stratification depends on the envelope of the pulse. Let us consider the two-parameter family of symmetrical initial profiles [28]:

$$q|_{\varepsilon=0} = \frac{1-\eta^2}{2} \ln \frac{1+\eta}{1-\eta} + (K-1)\,\eta + M\eta^3. \tag{4.19}$$

The solutions in the form (3.4) ($\delta_1\delta_2 = +1$) yield, that the definite parameters K and M values cause the steep decrease at any parts of the envelope, which leads to the stratification of the pulse. At the simple case, when the bend point formation occurs quickly ($\varepsilon^2 \ll 1$) near by the edges of the profile ($\eta^2 \to 1$) one may calculate the bend point coordinates (ε_k; η_k) with help of the condition (2.10). For instance, if $K = -12$; $M = 0{,}7$ we shall have: $\varepsilon_k = 0{,}45$; $\eta_k = \pm 0{,}9$. At the (τ, q) plane the coordinates of the bend point are: $\tau_k = -0{,}45$, $q_k = \pm 0{,}7$; the value of intensity at the bend point is $W_k = 0{,}25$. Fig. 7 illustrates the origin of symmetrical bend points at the pulse periphery and the beginning of the non-monotonous regime connected with growth of three maxima.
The process of decreasing curvature may lead to the origin of a new minimum instead of the maximum of envelope. For instance, in the case $\delta_1\delta_2 = -1$ the solution (3.11), describing the evolution of the initial profile (4.19), permits such self-stratification near by the centre ($\eta = 0$) (Fig. 8). In particular, when the parameters K and M values are $K = -1$, $M = -0{,}25$, then critical value ε is equal to $\varepsilon_k = 1/3$; ($\eta_k = 0$); the value of

intensity at the bend point is $W_k = 0{,}9$. The new minimum developing is accompanied by separation of two new maxima. After origin of the maxima their independent non-linear evolution may lead to the self-constriction and to the relative height growth. It is essential, that the above-mentioned self-stratification is not connected with the interference of the waves. Such interference may lead to "small-scale" distributions of

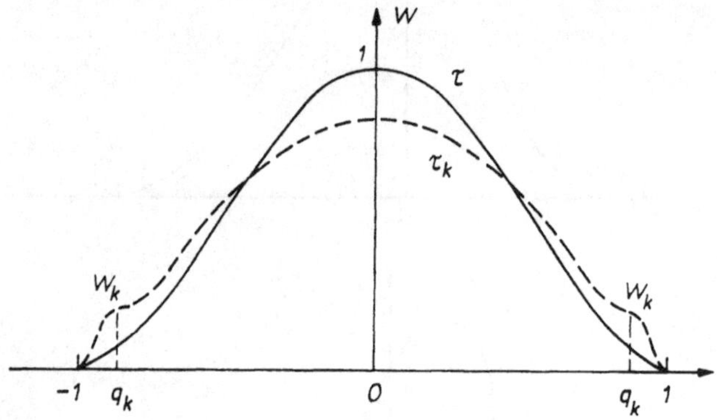

Fig. 7. The pulse periphery large-scale self-stratification (the continuous curve shows the initial profile, the dotted curve illustrates the new maxima origin)

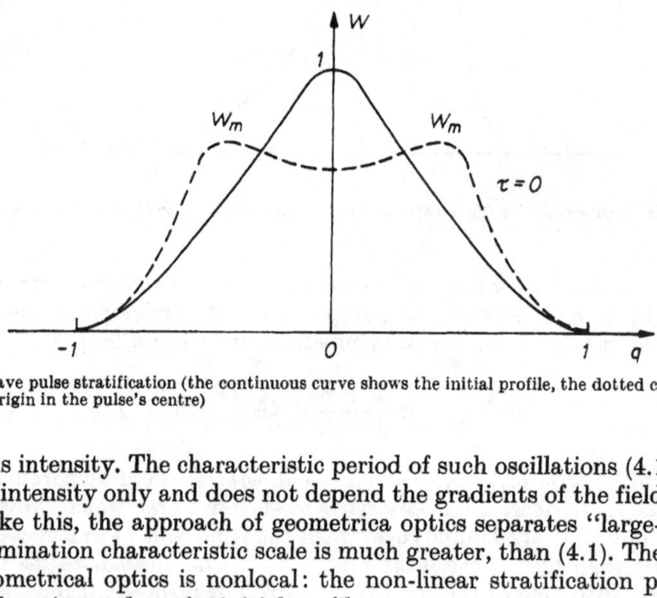

Fig. 8. The long wave pulse stratification (the continuous curve shows the initial profile, the dotted curve shows the new minimum origin in the pulse's centre)

the oscillations intensity. The characteristic period of such oscillations (4.1) depends the value of local intensity only and does not depend the gradients of the field and duration of pulse. Unlike this, the approach of geometrica optics separates "large-scale" oscillations. Such lamination characteristic scale is much greater, than (4.1). The stratification picture of geometrical optics is nonlocal: the non-linear stratification period depends on the pulse duration and on the initial profile.

4.4. The spectrum broadening stimulated by the intensity envelope deformation

The non-linear perturbations of phase velocity stimulate the development of phase distortions in the pulse. The phase shifts connected with the intensity envelope lead to the frequency envelope origin in the initially monochromatical pulse. The phase distortion takes place in the non-dispersive medium too [*20*], but in the dispersive medium the effects, accociated with the gradients of intensity are essential. The coupled deformation

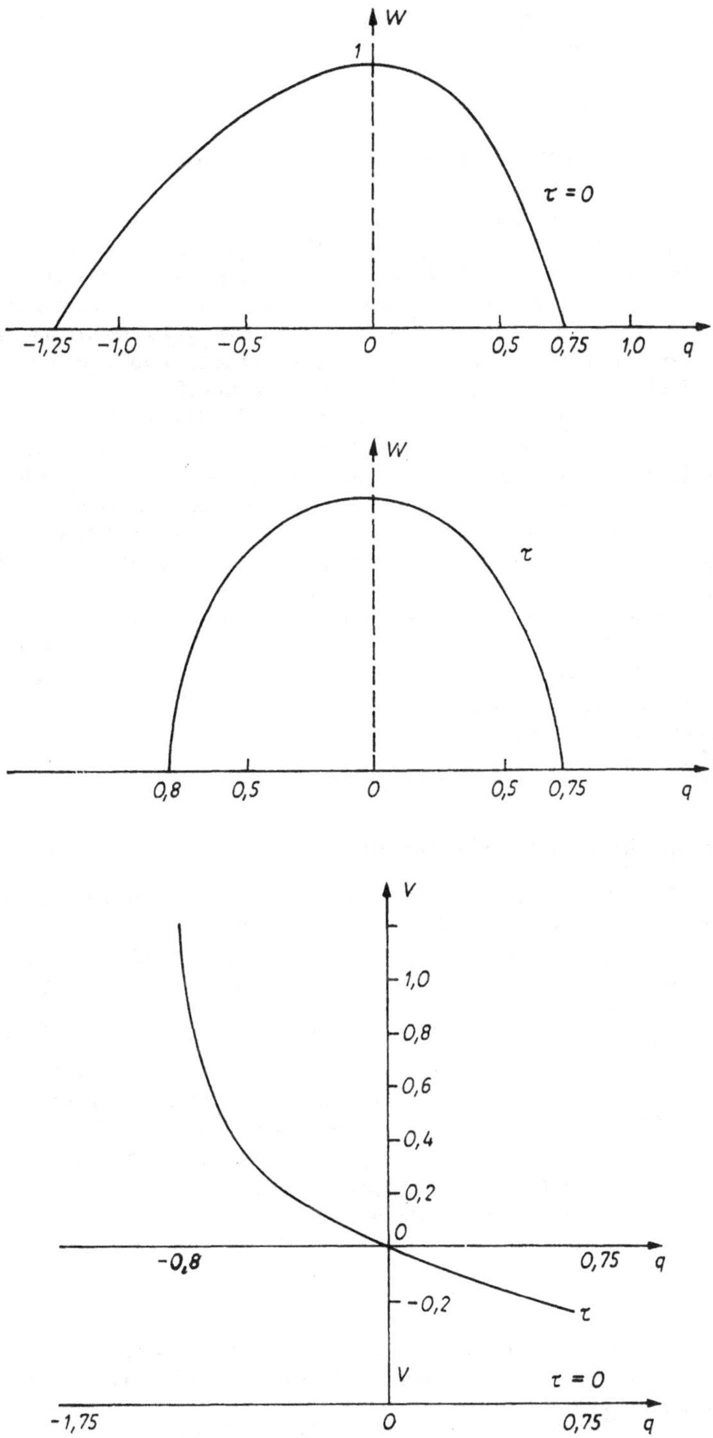

Fig. 9. The intensity and the frequency envelopes coupled non-linear deformation; fig. 9a — the initial profile; fig. 9b — the self-constriction of the pulse; fig. 9c — the broadening of the spectrum

of the intensity and the frequency envelopes is described by help of above-mentioned exact analytical solutions.
The development of the frequency envelope depends essentially on the type of growth of the pulse front curvature. Let us consider the one parameter family of the pulses with the abrupt fronts [42]:

$$q|_{\tau=0} = \eta - a\eta^2; \qquad v|_{\tau=0} = 0. \tag{4.20}$$

The envelope (4.20) corresponding to a value of parameter $a = -0{,}25$ is illustrated in the fig. 9a. The solution in the form (3.4) predicts the complex of the pulse (4.10) self-modulation phenomena development. Such complex contains the simultaneous processes of increase of pulse duration, the growth of pulse fronts curvature, the spectrum broadening. These results of evolution are illustrated at fig. 9b, corresponding to the case $\tau = 0{,}2$. The following evolution will stimulate the envelope shoke wave formation at the front of the pulse. Such singularity is satisfied to the condition (2.8).
The broadening of the spectrum increases together with the gradients of the intensity envelope. In case of abrupt fronts the maximum frequency broadening v develops at the pulse's edges (Fig. 9c). The sign of pulse frequency self-modulation is different at the ascending and the descending regions of the intensity envelope; near by the maximum of the pulse the function v is close to zero. Unlike this, in the case of the smooth edges of the pulse the frequency self-modulation is close to zero at the edges of the pulse too.
It is worth while to note the possibility of interaction between the initial frequency modulation of the pulse and the non-linear modulation development.
The dependence of the non-linear effect on the derivatives of intensity may effort or suppress the initial modulation. Addition of such effects may stimulate the considerable modulation or demodulation processes. The change of pulse modulation on the non-linear dispersive medium may become the effective source of information about the properties of this medium.

4.5. The "quadratic" approximation distortions in the non-linear waves theory

The "quadratic" model of non-linear permeability corresponds formally to the first term in the expansion of the function $\Delta\varepsilon(W)$ in a power series of W. However, the more realistic model contains the calculation of this approximation generalizations and the energy dissipation. Such generalization of non-linear geometrical equations (1.7), (1.10) may be fulfilled within the framework of transformation (2.3). This transformation does not impose any restrictions on the structure of the non-linear response of the medium $\Delta\varepsilon(W)$. For the different forms of the function $\Delta\varepsilon(W)$ application of the transformation (2.3) to the eikonal equation (1.10) leads to a linear differential equation (1.10) in which the function $\Delta\varepsilon(W)$ is a coefficient. Such a transformation is especially convenient for an analysis of the non-linear wave processes in media, characterized by a complex form of the response $\Delta\varepsilon(W)$. For instance, such approach may be utilized to the problem of self-action of waves in a medium with saturation of the non-linearity [34]. Such phenomena may be described with help of the "exponential" model [35], [37]

$$\Delta\varepsilon(W) = 1 - \exp(-\beta W). \tag{4.21}$$

Such approximations, connected with the Boltzmann's type formulae, are convenient in the non-linear problems of plasma radio-physics [30], [39], [41]. Another approximation has the "fraction" type:

$$\Delta\varepsilon(W) = \frac{\beta W}{1 + \gamma W}. \tag{4.22}$$

In the case of weak intensities both models (4.21) and (4.22) transfer to the "quadratic" model.

The self-action of waves in the medium with the saturation of the non-linearity was analysed in [*34*] within the framework of eq. (2.4) and the mode (4.22). The role of saturation was illustrated in [*34*] with help of the one-parameter family of pulses with different curvature of the front (A is the parameter):

$$W|_{\tau=0} = \frac{1}{\gamma}\left[\left(\frac{q}{A}\right)^{-1/2} - 1\right]. \tag{4.23}$$

The non-linear dynamics of pulses (4.23), described within the framework of exact analytical solutions of eq. (2.4), is illustrated in Fig. 10. Such self-action contains the simultaneous development of the processes of the pulse-self-compression, the formation of it's front, and the pulse auto-modulation. Fig. 10 shows, that near by the edge of the

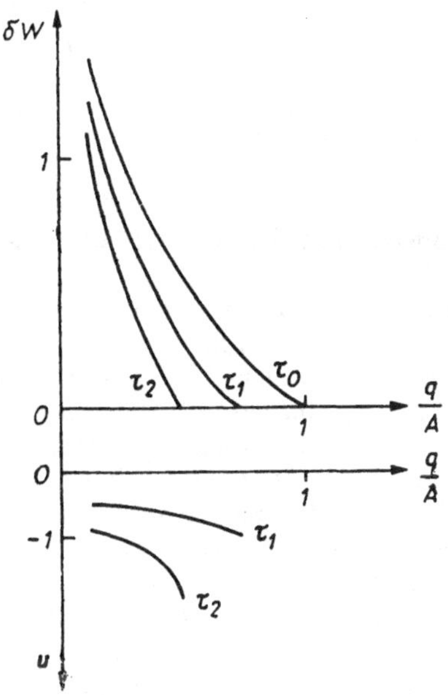

Fig. 10. Localization of self-modulated region inside a medium with saturation of the non-linearity

pulse, when the intensity is relatively weak ($|\gamma W| \ll 1$), the slope of the front increases and the modulation grows proportional to the path traversed. This solution limit coincides with the corresponding result of the "quadratic" model. However, in contrast to the "quadratic" model, the saturation model (4.22) described the gradual weakening of the front formation and auto-modulation non-linear effects when the pulse passes in the saturation region for large values of the intensity. These results illustrate the interesting effect of localization of the auto-modulation region inside of a medium with saturation of the non-linearity. The greatest effects is concentrated in the region with considerable values of the intensity gradient.

The linear dissipation distorts the values of the gradient of intensity. Such distortions may become an additional source of the auto-modulation in the non-linear medium. This additional modulation was analysed in [*34*] in connection with the evolution of the pulse $W|_{\tau=0} = Bq$, where $B =$ constant. The corresponding solution is valid both in the dissipative and in the active media. In the active medium slope of the pulse front and it's modulation, unlike the dissipative case, will increase inside the medium. So, the non-

linear response different approximation expressions and the waves linear absorption may be included to the general scheme of non-linear geometrical optics.
The change of the pulse gradient inside the medium leads to a rather complicated form of the distributions of solution-like waves. In the transparent medium the non-linear permeability saturation (4.21), the solitary waves, and the non-linear periodical regimes were analysed in [36]. Another "quadratic" model distortion, connected with the "power-law" model in the form

$$\Delta\varepsilon(W) = \beta W + \gamma W^2 \tag{4.24}$$

leads [34] to the generalization of the trivial "soliton" regime, when the waves distribution intensity is finite even at the limits $q \to \pm\infty$. The intensity values of at the infinity depend on the parameter γ in (4.24).
The above-mentioned solutions of geometrical optics illustrate the dependence of the non-linear evolution of the pulse on the envelopes both at the region of maximal intensity and at the periphery. In this connection it is worth while to note, that the approximative consideration of the region of maximal intensity only and neglection of the periphery parts may lead to wrong results. During the process of non-linear evolution the change of profile curvature may stimulate the deformation of the essential pulse structure both at the centre and at the periphery.

5. The large-scale rebuilding of the wave fields with the complicated geometry

The cylindrical wave non-linear evolution stationary picture is described within the framework of equations (1.7), (1.10) of non-linear geometrical optics. Here the function v describes the distribution, connected with the non-linear curvature of the wave front. Variables τ and q signify the longitudinal and the transversal co-ordinates resp. This analogy permits to analyse the cylindrical wave beams, bounded on entry into the non-linear medium by a slit or an opaque strip with help of exact analytical solutions, constructed below (§ 4). Unfortunately the analyse of the three-dimensional non-linear wave beam evolution is rather more difficult. Therefore knowledge of such general tendencies of evolution becomes very important. Such investigation to tendencies within framework of asymptotic [32], numerical [22], variation [38] and qualitative [53] methods illustrated the points of view existence, connected with the wave-guide [9] regime and the "focus" or "many-focus" structure [22], correspondingly. Thus, the series of problems, associated with the three-dimensional localized wave non-linear evolution, still remain. The details of evolution, essential in both wave-guide and "focus" conceptions, are of special interest. In both cases the stationary self-action picture in the region, adjointed to the boundary of non-linear medium, depends on the initial distributions of intensity and direction of rays. Therefore, it is useful to analyse systematically the large-scale field structure picture at this primary evoluting stage. The equations of non-linear geometrical optics, corresponding to this stage, have the form:

$$\frac{\partial W}{\partial \tau} + \operatorname{div}_\perp (W\boldsymbol{v}) = 0; \tag{5.1}$$

$$\frac{\partial \boldsymbol{v}}{\partial \tau} + (\boldsymbol{V}\nabla)\,\boldsymbol{V} - \delta_1\delta_2 \cdot \beta \cdot \nabla_\perp W = 0. \tag{5.2}$$

Here $\delta_1\delta_2 = \pm 1$ values correspond to the focussing and defocussing problems.
For simplicity, the model of non-linear permeability in (5.2) is supposed to be "quadratical". By analogy to the two-dimensional problem one may suppose that the solutions of

the set (5.1), (5.2) are very sensitive to the initial distribution of localized waves. But, unlike the two-dimensional problem, the transformations of Legendre type (2.3) or (2.5) do not reduce the equations of the three-dimensional (5.1), (5.2) to the linear system. The transformation to the inverse functions, like (2.1), (2.2), is useless now, and the direct solutions, of (5.1), (5.2) set are desirable. Some progress in this problem is stimulated by the approach of automodel solutions.

5.1. The automodel solutions in the dynamics of anisotropic wave distributions

Let us consider the three-dimensional intensity and phase distributions in the power wave beam without the axial symmetry. The analyse of the anisotropic Langmuir waves self-action dynamics [*33*], the transversal electromagnetic wave in the non-linear dielectric [*51*] and the laser beam two coaxial modes cross focusing [*49*] within the framework of the numerical solution illustrate the possibility of such distributions asymmetry increase in the non-linear medium. It is convenient to analyse such self-actions the main tendencies with the help of two-parameter distributions family. These parameters are connected with the initial intensity and phase distributions anisotropy. Let us suppose that the constant intensity curves in the initial field distribution have the elliptic forms: (τ is the beam propagation direction)

$$W|_{\tau=0} = \left.\begin{matrix} 1 - \dfrac{x^2}{a^2} + \dfrac{y^2}{b^2} \\ 0 \end{matrix}\right\} \begin{matrix} x^2 \leqq a^2; & y^2 \leqq b^2; \\ x^2 \geqq a^2; & y^2 \geqq b^2; \end{matrix} \quad a^2 \leqq b^2. \tag{5.3}$$

Then, the initial phase front of the narrow beam may be represented in form of the ellipsoid surface part:

$$S|_{\tau=0} = \frac{x^2}{2R_1} + \frac{y^2}{2R_2}. \tag{5.4}$$

Here x and y are the orthogonal axes at the boundary plane ($\tau = 0$), and R_1, R_2 are the phase front curvature radii near the maximum of intensity at the point $x = 0$, $y = 0$; a and b are the beam width initial values at the directions x and y resp. The initial distributions family (5.3), (5.4) is characterized by two parameters:

$$\lambda = \frac{R_1}{R_2}; \qquad \mu = \frac{a^2}{b^2} \leqq 1.$$

The exact analytical solution of the set (5.1), (5.2), satisfying to the boundary conditions (5.3), (5.4), may be represented in the automodel form [*50*]:

$$W(x; y; \tau) = \frac{1}{f_1(\tau) \cdot f_2(\tau)} \left[1 - \frac{x^2}{a^2 \cdot f_1^2(\tau)} - \frac{y^2}{b^2 \cdot f_2^2(\tau)}\right]; \tag{5.5}$$

$$S = \frac{\gamma_1(\tau) \cdot x^2}{2} + \frac{\gamma_2(\tau) \cdot y^2}{2}; \qquad \boldsymbol{u} = \nabla_\perp s. \tag{5.6}$$

Here the dimensionless functions $f_1(\tau)$, $f_2(\tau)$, $\gamma_1(\tau)$, $\gamma_2(\tau)$ describe the variable beam width and phase front curvature resp. It is essential, that the beam remains his "elliptic" structure with the variable eccentricity.
It is useful to pass new variables

$$t = \tau \cdot (2\beta)^{-1/2} \cdot a^{-1} \cdot \mu^{-1/4}; \qquad \xi = f_1; \qquad \eta = f_2 \cdot \mu^{-1/4}. \tag{5.7}$$

Let us substitute (5.5)—(5.7) into (5.1)—(5.2); we obtain, that the values of beam width are governed by the following equations:

$$\frac{d^2\xi}{dt^2} = -\frac{\delta_1\delta_2}{\xi^2\eta}; \qquad \frac{d^2\eta}{dt^2} = -\frac{\delta_1\delta_2}{\xi\eta^2}. \tag{5.8}$$

The non-linear system (5.8) has two conservation laws:

$$\frac{1}{2}\left(\frac{d\xi}{dt}\right)^2 + \frac{1}{2}\left(\frac{d\eta}{dt}\right)^2 - \frac{\delta_1\delta_2}{\varepsilon\eta} = \varepsilon_0, \tag{5.9}$$

$$\xi^2 + \delta_1\delta_2\eta^2 = p\cdot t^2 + m\cdot t + c. \tag{5.10}$$

Here the constants ε_0, p, m, c are determined by the initial conditions. It is convenient to use the mechanical analogy and suppose the functions $\xi(t)$ and $\eta(t)$ as any moving point co-ordinates at the $(\xi; \eta)$ plane and the variable t as the time. In such approach the eq. (5.10) describes the trajectory of the motion, and eq. (5.9) describes the energy conservation law, associated with this potential motion ($-\delta_1\delta_2(\xi\eta)^{-1}$ is the "potential").
The formulae (5.9), (5.10) are just both in the focussing ($\delta_1\delta_2 = +1$) and defocussing ($\delta_1\delta_2 = -1$) media. Let us consider, for simplicity, the focussing medium. In this case the substitution

$$\xi = R\cdot\cos\varphi; \qquad \eta = R\cdot\sin\varphi; \tag{5.11}$$

permits to derive from the eq. (5.9), (5.10) the evident expressions for the functions $R(t)$, $\varphi(t)$. These solutions describe the beam cross section constriction into a narrow extensive region. The initial divergence of rays and the self-action process competition show the possibility of two anisotropic beam self-focussing regimes.
The monotonous regime is illustrated in Fig. 11 and Fig. 12. Unlike this, the non-monotonous regime is shown in Fig. 13. In this case the narrow waves localization region development is accompanied with oscillations of the cross-section.

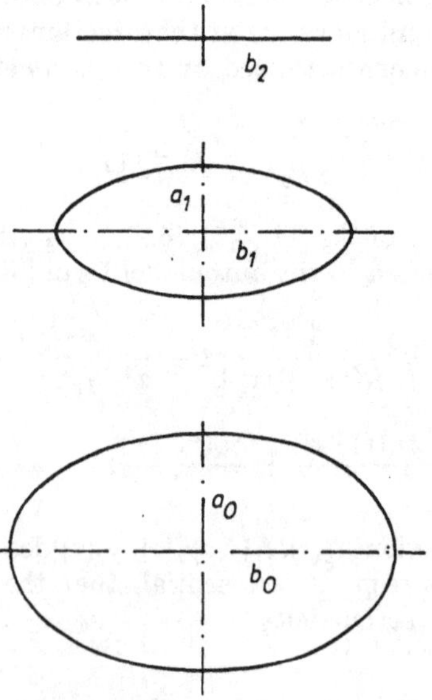

Fig. 11. The monotonous evolution of a plane anisotropic beam ($a\cdot b^{-1} = 1/3$); cross section

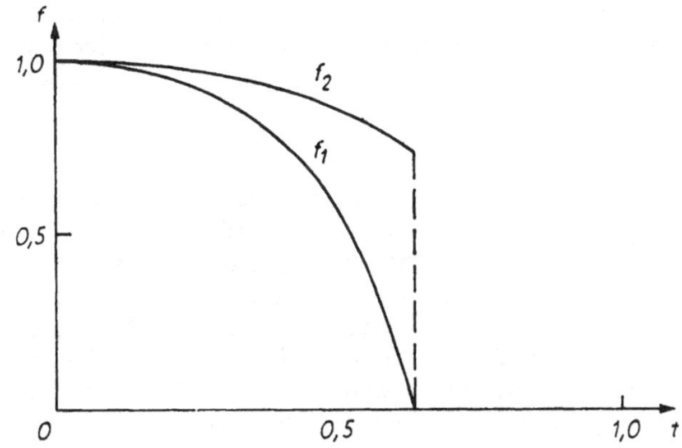

Fig. 12. The decrease of the "elliptic" beam width values during the self-focusing process

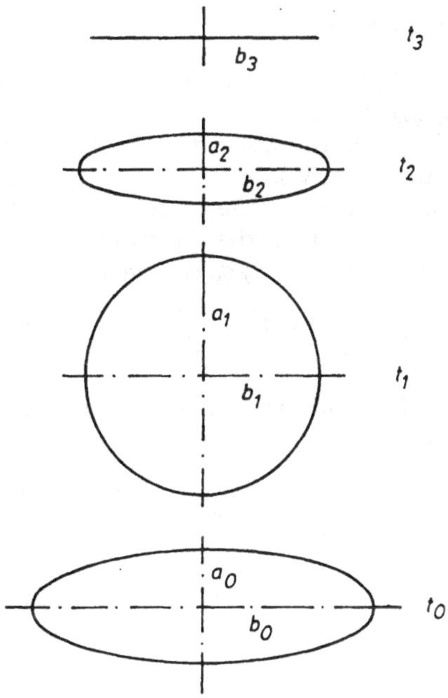

Fig. 13. The non-monotonous regime of elliptical distribution rebuilding

The private case $\mu = \lambda = 1$ corresponds to the axial-symmetrical parabolic wave beam with the spherical phase front:

$$W|_{\tau=0} = 1 - \frac{\varrho^2}{a^2}; \qquad v|_{\tau=0} = \varrho \cdot \theta; \qquad x^2 + y^2 = \varrho^2. \tag{5.12}$$

Such evolution of beam leads to the focussing of beam into one point — the "focus" [26]. The profiles (5.5), (5.12) evolution in the "quadratic" non-linear permeability model is described by the automodel solutions of the set (5.1), (5.2). Such solutions predict the origin of singularities connected, with the intersection of rays near by the maximum of intensity inside the focussing medium. In the defocussing medium the "elliptic" distribution (5.5) expansion will be accompanied with it's eccentricity decrease [41].

5.2. The non-linear aberrations of the wave beam

The collaps of the parabolic beam (5.12) may be compared with the focussing action of the lens. By analogy, the "elliptic" beam self-focussing is similar to the astigmatic lens effect. Such idealized models predict the crossing of beam rays into one point-"focus" or into the segment. However, the real models of optics, dealing with the lens surface curvature changes and the beam angular divergence considerable values, predict the series of aberrations. Just the same factors, connected with the profiles (5.12) distortions stimulate the non-linear aberrations development. The theory of non-linear aberrations of the different order may be developed as the perturbations theory, associated with the profiles, close to the parabolic one (5.12). Such evolution of profiles is supposed to be described with the expressions in the (W, v) space:

$$\varrho = \varrho_0(W, v) + \varrho_1(W, v); \qquad t = t_0(W, v) + t_1(W, v). \tag{5.13}$$

Here ϱ_0 and t_0 correspond to the "parabolic" solution, and ϱ_1 and t_1 are the perturbations. These functions ϱ_1 and t_1 are governed with the equations, derived from the master system (5.1), (5.2) after it's Legendre transformation. It is essential, that such equations in the (W, v) space are linear in each order of the perturbation theory.
Because of this it is possible to generalize the analytical technique of the two-dimensional problem and to utilize this technique in the theory of three-dimensional perturbations. It is convenient to introduce in the (W, v) space the spheroidal coordinates through the formulae [48]

$$W = (1 + \varepsilon^2)(1 - \eta^2); \qquad v = \sqrt{2}\,\varepsilon\eta$$

(the plane wave in the focusing medium is considered here). The parabolic solution has in the $(\varepsilon; \eta)$ space the form:

$$\varrho_0 = -\frac{\eta}{\sqrt{1 + \varepsilon^2}}; \qquad t_0 = -\frac{\varepsilon}{\sqrt{2}} \cdot \frac{1}{\sqrt{1 + \varepsilon^2}}. \tag{5.14}$$

By analogy to the two-dimensional problem the perturbations ϱ_1 and t_1 are determined by the formulae:

$$\varrho_1 = -\frac{1}{\sqrt{2}\,(\varepsilon^2 + \eta^2)} \left[\varepsilon(1 - 2\eta^2)\frac{\partial\psi}{\partial\eta} + \eta(1 + 2\varepsilon^2)\frac{\partial\psi}{\partial\varepsilon}\right], \tag{5.15}$$

$$t_1 = \frac{1}{2(\varepsilon^2 + \eta^2)}\left(\eta\frac{\partial\psi}{\partial\eta} - \varepsilon\frac{\partial\psi}{\partial\varepsilon}\right). \tag{5.16}$$

The function ψ describing the perturbations of the axial-symmetrical distribution is represented by the expansion:

$$\psi = \sum_{m=0} \{P_m(\eta)\,[A_m f_m^{(1)}(\varepsilon) + B_m f_2^{(2)}(\varepsilon)] + Q_m(\eta)\,[C_m f_m^{(1)}(\varepsilon) + D_m f_m^{(2)}(\varepsilon)]. \tag{5.17}$$

The structure of formula (5.17) is similar to the corresponding expression (3.4) in the two-dimensional theory, however, the other eigen-functions are utilized in the three-dimensional problem [48]:

$$P_0 = 1, \qquad Q_0 = \frac{1}{2}\ln\frac{1-\eta^2}{\eta^2}, \qquad f_m^{(1)} = \mathrm{ch}\,(K \cdot \mathrm{arsh}\,\varepsilon),$$

$$P_1 = 1 - 2\eta^2, \qquad Q_1 = P_1 \cdot Q_0 - P_0, \qquad f_m^{(2)} = \mathrm{sh}\,(K \cdot \mathrm{arsh}\,\varepsilon), \tag{5.18}$$

$$P_2 = 1 - 6\eta^2 + 6\eta^4, \qquad Q_2 = P_2 \cdot Q_0 - \frac{3}{2}P_1 .$$

Here P_m are the Jacobi polynomials $K = \sqrt{2m(m+1)}$ $m = 1; 2; 3\ldots$ In the case $m = 0$ the functions f_0 and $f_0^{(2)}$ have the form:

$$f_0^{(1)} = 1; \qquad f_0^{(2)} = \ln\left(1 + \sqrt{1+\varepsilon^2}\right).$$

The perturbations of the initial parabolic profile are determined by eq. (5.15) at the boundary plane $\varepsilon = 0$. The "smooth" distribution coincides with the parabolic one for $\varrho = 0$ and $\varrho = 1$ and differs slightly from it for other values of ϱ $(0 < \varrho < 1)$ may be described with the help of the functions $Q_n(\eta)$ (5.18). The "abrupt" profile edges may be expressed with the help of the functions $P_n(\eta)$ (5.18) [41].

The deviations from the parabolic shape are rather small for small values of variable ε. However, when $\varepsilon \gg 1$, near by the region of high intensity, those deviations are rather considerable. The development of initial perturbations leads to the origin of singularities; the positions of these singularities are determined by formulae (2.7), (2.9).

It is essential, that such aberrations restrict intensity values; unlike the parabolic profile, the intensity in the singular point has finite values. Moreover, the above-mentioned procedure automatically leads to finite intensity values in the higher orders of the theory of perturbations. The axial symmetric intensity of the beam in the approximation of geometrical optics increases 15—20 times. After the aberration beginning the multiray domains are formed. The equations of geometrical optics do not suit to this case. The other beam evolution tendency is connected with the curvature phase front partial decrease. This process leads to the beam lamination; such situation is possible in the defocussing medium too [47]. The singularities of geometrical optics determine the region of large field gradients inside the non-linear medium. The following wave evolution is connected with the crossing of waves and with the interference phenomena.

The dissipation of linear waves cannot distort qualitatively these tendencies display if the dissipation process is weak enough [15]. Unlike this, the many-photon dissipation distorts the region of intensity maximum mainly; because of this, the peculiar effect of many-photon dissipation and initial parabolic distortion mutual compensation is possible [47].

Conclusion

It is worth while to note now the main achievement and difficulties of the non-linear geometrical optics.

1) The new approach in the non-linear theory of waves in the dispersive medium is developed. The non-stationary large-scale picture of the localized waves field non-linear evolution is described in the framework of such approach by the wide classes of exact analytical solutions of the self-action equation. The special non-linear wave equation transformation reduced the wave self-action problem to the linear one in the intensity-modulation space. The corresponding solutions illustrate the non-linear pulse dynamics tendencies, based on the initial intensity-phase distributions.

2) The exact solutions of non-linear geometrical optics have very simple form, represented by the eigen-functions of the Laplace equation in the above-mentioned intensity-modulation space. Such representation permits to divide the qualitatively different effects of the non-linear deformation of the pulse, to unify a great number of the analytic investigation technique of pulses, to generalize the geometrical optics approach to the complicated problems of non-linear permeability.

3) The non-stationary large-scale pulse evolution in the dispersive medium stimulates the pulse envelope slope growth or decrease at any localized wave distribution part. The first tendency development leads to the quick intensity increase area origin in the initially smooth waves distribution. The other tendency leads to the horizontal beam point formation at the intensity profile and to the following pulse large-scale stratification. This self-stratification does not depend on the wave-length, but depends on the initial pulse duration and the intensity envelope. Such characteristical points of wave evolution are represented by the singularities of the exact solutions of non-linear geometrical optics. The initial distribution of pulse gradients may stimulate these origin of singular points both near by the intensity maximum and at the periphery of the pulse.

The non-linear geometrical optics achievements permit to outline the following fields of such theory application and development:

1) The non-linear waves dynamics modelling

The wave field evolution in the non-linear dispersive medium is described in this review with help of the dimensionless variables q and τ and the dimensionless functions W and v in the framework of the approximation of geometrical optics. Such wording permits to compare one and the same evolution tendencies in the fields of the different physical nature, described by the non-linear Schrödinger equation (1.5). In particular, the large-scale localized Langmuir waves [*33*], ion-acoustic waves [*10*] and high-frequency electromagnetic waves [*34*] pulse dynamics in plasma is described with help of the comparable parameters. This analogy is observed in the several cases [*10*] of the electro-magnetic wave branches in the magnetoactive plasma. Another utilizations of the non-linear-Schrödinger equation lead to similar comparisons too. Such similarity shows the possibility of the modelling of the one wave branch evolution with help of another wave branch re-building picture. This analogy may be especially fruitful, if one of these branches is reproduced well in optical or radiophysical experiments.

2) The many-dimensional non-linear geometrical optics problems

These problem sare very difficult, because the exact analytical solutions (§ 3) are constructed in the two-dimensional case only (one-stationary plane waves evolution and the stationary cylindrical waves re-building). In the other cases the Legendre transformations of equations of non-linear geometrical optics are not valid. Such difficulties are aggravated, if the wave beam propagates through the non-stationary or non-homogeneous medium, because the corresponding phase and amplitude corrections are added to the non-linear perturbations. Such wave beam dynamics estimations are limited usually by the framework of the theory of perturbations. This important circle of problems demands the essential development of analytic technique.

3) The geometrical optics and the wave stages of the "large-scale" pulse non-linear evolution

The role of diffraction phenomena differs essentially at the localized field non-linear evolution different stages. During the "geometric optics" stage of the "large-scale" wave distribution the diffraction phenomena form the field "fine structure" on the background, connected with the intensity and phase smooth envelopes. The following non-linear evolution process may emphasize such "fine-scale" non-homogeneties more relief. Namely, such non-homogeneties may limit the energy cumulation at the self-construction process of the pulse. The diffractive and the non-linear wave distortions mutual compensation fixes the stationary wave localization region — the plane waves soliton and the three-dimensional non-linear wave-guide [*53*]. Unlike such compensation possibility, the attemps of diffractive and non-linear non-stationary phenomena simultaneous calculation, demand the exact self-consistent solutions of eq. (1.5) construction. Following the ideas of V. E. Zakhazov and A. B. Shabad [*25*] it is useful to connect the construction of such solutions with the inverse scatttering problem, associated with the Schrödinger equation. The principal result in the framework of such approach was obtained by S. V. Manakov [*23*] in connection with the Fraunhofer diffraction problem in the non-linear medium. Namely, the classical formula of Fraunhofer maxima directions θ was generalized in [*23*] in the Form:

$$\theta^2 = \theta^2_m + 2n_1 |E|^2 = \left(\frac{2\pi m}{kl}\right)^2 - 2n_1 |E|^2. \tag{5.19}$$

Here E is the amplitude of the wave, l is the Fraunhofer lattice split width, K is the wave number, m is the diffraction maximum order, $n_1 |E|^2$ is the non-linear part of the refractive index. It is essential, that the linear theory predicts the diffractive picture diffusion far from the lattice. Unlike this, the non-linear theory predicts the origin of the particular intensive field periodical structure inside the medium even far from the Fraunhofer lattice [*23*]. Such structure origin has the threshold character; the value of threshold amplitude, corresponding to the zero order maximum, is calculated from (5.19), when $\theta = 0$.

This result describes the diffraction of the plane phase front with the constant intensity, the corresponding initial pulse has the rectangular form. The generalization of this important result, connected with the variable intensity profiles, represents the very actual problem. Unlike the Fraunhofer diffraction, the non-linear phase shift in the different points of phase front is changed. By analogy to the non-linear Fraunhofer diffraction it is convenient to discuss the "non-linear phase lattice" or the "non-linear phase screen" in this case. The analyse of the "non-linear phase lattice" effect in the framework of the inverse scattering problem approach will be especially useful. These results permit to compare the geometrical optics and the wave optics exact analytical solutions in the non-linear medium for one and the same intensity envelopes. Such comparison will help to estimate the accuracy of the approximation of geometrical optics. This task represents one of the aspects of the important theoretical problem, associated with the limit passage from the wave descriptin to the geometrical optics picture. Such passage connected with the decrease of wave length, gives rise to many difficulties even in the linear theory. The corresponding non-linear problem seems to be much more complicated. The non-linear geometrical optics began to form in connection with the analyse of non-linear wave processes during the recent last years. The first successes of this approach in radio-physics and optics stimulate the adjacent problems development in acoustics, plasma physics, and wave electronics. This simple method is useful for peculiar "non-linear intuition" promotion in the different fields of physics. One may hope, that the general interest to the waves theory in the non-linear distributed systems will stimulate the following progress of the effective method of non-linear geometrical optics.

Acknowledgements

The autor is very grateful to Professor B. Wilhelmi for his encouragement, connected with this review preparation. The author thanks Professor E. P. Velikhov, Professor V. S. Letokhov, Professor V. V. Migulin for their kindful criticism and fruitful discussions.

References

[1] M. SCHUBERT, B. WILHELMI, „Einführung in die nichtlineare Optik", Leipzig (1971).
[2] G. B. WHITHAM, "Linear and Non-Linear Waves", Mir (1977).
[3] N. BLOEMBERGEN, "The Non-Linear Optics", Mir (1965).
[4] V. L. GINZBURG, "The propagation of the electromagnetic waves in Plasma", Nauka (1967).
[5] V. P. SILIN, "The parametrical action of the high power radiation on the plasma", Nauka (1973).
[6] G. A. ASKARYAN, Zh. eksper. teor. Fiz. **42**, 1567 (1967).
[7] S. A. ACKMANOV, A. P. SYCKORUKOV, R. V. CHOKHLOV, Soviet Physics-Uspechi **93**, 191 (1967).
[8] O. V. RUDENKO, S. I. SOLYAN, "The theoretical foundation of the non-linear acoustics", Nauka (1975).
[9] G. A. ASKARYAN, Zh. eksper. teor. Fiz. Pisma **4**, 144 (1966).
[10] M. KAKO, Supplement Progr. theor. Phys. **55**, 120 (1975).
[11] K. WATANABE, T. TANIUTI, Report Dept. Phys. Nagoya Univ., DPNV-83-76 Oct. 1976.
[12] T. TSUZUKI, J. Low Temper. Phys. **4**, 441 (1971).
[13] M. TODA, Supplement Progr. theor. Phys. **45**, 174 (1970).
[14] P. K. SHUKLA, M. Y. YU, K. H. SPATSCHEK, Planet Space Sci. **24**, 1093 (1977).
[15] L. STENFLO, J. Fluid Mech. **18**, 472 (1977).
[16] R. Y. CHIAO, E. GARMAIRE, C. H. TOWNES, Phys. Rev. Letters **13**, 479 (1964).
[17] A. V. GAPONOV, M. A. MILLER, Zh. eksper. teor. Fiz. **34**, 242 (1958).
[18] P. LALLEMAND, N. BLOEMBERGEN, Phys. Rev. Letters **15**, 1010 (1965).
[19] J. B. ZELDOVICH, Y. P. RAUZER Zh. eksper. teor. Fiz. Pisma **3**, 37 (1966).
[20] F. DE MATRINI, C. H. TOWNES, T. K. GUSTAFSON, P. L. KELLEY, Phys. Rev. **164**, 312 (1967).
[21] L. A. OSTROVSKY, Zh. eksper. teor. Fiz. **51** 1189 (1966).
[22] V. N. LYGOVOI, A. M. PROKHOROV "Soviet Physics-Uspechi" **111**, 94 (1973).
[23] S. V. MANAKOV, Zh. eksper. teor. Fiz. **65**, 1392 (1973).
[24] M. J. LIGHTHILL, Proc. Roy. Soc. **299 A**, 28 (1967).
[25] V. E. ZAKHAROV, A. B. SHABAD, Zh. eksper. teor. Fiz. **61**, 118 (1971).
[26] V. I. TALANOV, Radio-Physics **7**, 764 (1964).
[27] L. D. LANDAU, E. M. LIFSHITZ, "The Hydrodynamics", Nauka (1972).
[28] A. B. SHVARTSBURG, Zh. eksper. teor. Fiz. **66**, 920 (1974).
[29] V. V. KOROBKIN, A. M. PROKHOROV, Zh. eksper. teor. Fiz. Pisma, **12**, 216 (1970).
[30] A. V. GAPONOV, L. A. OSTROVSKY, M. Y. RABINOVICH, Radio-Physics **13**, 163 (1970).
[31] V. I. BESPALOV, V. I. TALANOV, Zh. eksper. teor. Fiz. Pisma, **3**, 471 (1966).
[32] V. E. ZAKHAROV, Zh. eksper. teor. Fiz. **64**, 1627 (1973).
[33] V. E. ZAKHAROV, Zh. eksper. teor. Fiz. **62**, 5 (1972).
[34] A. B. SHVARTSBURG, Zh. eksper. teor. Fiz. **70**, 967 (1976).
[35] V. Y. KOLOKOLOV, Radio Physics XVIII, 1594 (1975).
[36] L. M. GORBUNOV, Soviet Physics-Uspechi **109**, 631 (1973).
[37] A. B. SHVARTSBURG, Doklady AN USSR **200**, 575 (1971).
[38] J. M. CRICHEVER, Soviet Mathematics-Uspechi **6**, 183 (1977).
[39] A. B. SHVARTSBURG, Opto-Electronics **8**, 393 (1976).
[40] V. I. TALANOV, Zh. eksper. teor. Fiz. Pisma, **2**, 218 (1965).
[41] A. B. SHVARTSBURG, "The geometric optics in the non-linear waves theory", Nauka (1976).
[42] A. B. SHVARTSBURG, Phys. Letters **59 A**, 208 (1974).
[43] V. J. EMERY, J. Math. Phys. **11**, 1893 (1970).
[44] A. B. SHVARTSBURG, Phys. Letters **48 A**, 257 (1974).
[45] G. CHEW, M. COLDBERGER, F. LOW, Proc. Roy. Soc. **236 A**, 112 (1956).
[46] G. S. GARDNER, J. GREEN, M. KRUSKAL, K. MIURA, Phys. Rev. Letters **19**, 1095 (1967).

[47] A. B. SHVARTSBURG, Radio-Physics XIX, 2 (1976).
[48] A. B. SHVARTSBURG, Optics Communications X, 127 (1974).
[49] M. S. SODHA, V. K. TRIPATHI, V. P. NAYYAR, Optics Communications IX, 381 (1972).
[50] A. B. SHVARTSBURG, Radio-Physics XIII, 775 (1970).
[51] V. V. VOROB'EV, Radio-Physics XIII, 1905 (1970).
[52] M. MOORE, Opto-Electronics **5**, 189 (1973).
[53] V. M. ELEONSKY, L. G. OGANESYAN, V. P. SILIN, Zh. eksper. teor. Fiz. **63**, 532 (1972).
[54] A. N. TICHONOV, Soviet Physics-Uspechi **113**, 127 (1974).

Fortschritte der Physik 28, 35—65 (1980)

Space-Time Approach to the Description of Cumulative-Type Processes

B. N. KALINKIN, A. V. CHERBU[1]), V. L. SHMONIN[2])

Joint Institute for Nuclear Research, Laboratory of Theoretical Physics, Dubna, USSR

Abstract

An approach to the description of cumulative-like processes is developed. It explicitly uses space-time factors determining the development of these processes and some general features of hadronic inclusive spectra. The results are compared with all the available experimental data. The predictions testable in current and planned experiments are given.

Contents

---

[1]) Institute of Applied Physics, Academy of Sciences MSSR, Kishinev, USSR.
[2]) Institute of High Energy Physics, Academy of Sciences Kaz. SSR, Alma-Ata, USSR.

1. Introduction

In recent years the investigations in one of the most important fields of high-energy physics, relativistic nuclear physics, have been revived.
Besides traditional general problems of multiple production in matter, quite special and rather interesting problems concerning the description of the inclusive spectrum of produced particles at magnitudes of kinematical variables, unattainable in hadron-hadron collisions (e.g., particles with large momenta flying into the back hemisphere) have occupied a noticeable place in it.
A. M. Baldin and the group headed by V. S. Stavinsky [1] in JINR have brought a very ponderable contribution into the studying of these problems. The circle of interesting experiments was carried out by Leksin's group [2] in JTEP.
We are also to note the works of american physisists. Extremely great resonance was caused by works by Cronin [3, 4] and co-workers.
Since last years a number of other works have appeared (see, e.g., [5, 6]).
During a short period of time these studies acquire wide popularity and cause sharp polemics.
The popularity is explained by a great interest in the problem and by hopes to extract a new additional information when solving.
The polemics (as in general, in interpreting multiple production in matter) has resulted from the clash of two points of view — simplified approaches (e.g., application of the rescattering mechanism or taking into account of Fermi's motion in a nuclear system) and the opinion that we deal with the manifestation of hadron collective interaction.
In this paper we attempt to represent quite completely the picture of cumulative processes we develop and its basic results [7—12]. Here we shall not describe in detail other approaches, as it is impossible within the chosen scheme. Such an approach has, generally speaking, another reason — unfortunately, the model in question and its capabilities are not well known.
Nevertheless, it is the semiphenomenological limit of the approach the realization of which in other, not less important processes (multiple production in the hadron-nucleus interaction [13], a nucleus-nucleus one [14], and, recently, that of hadron-hadron [15]) leads to rather good results, and the agreement with experiment is not the only reason: of most importance is the possibility to establish, in a number of points, qualitative and quantitative connection of the approach with modern concepts on hadron structure in the quark-glue model and in the framework of quantum chromodynamics.
Before speaking on the space-time approach of cumulative processes, we shall make one more general comment.
We are inclined to think that cumulation is a manifestation of a hadron collective interaction, and it is probably a way to a new information. Nevertheless, we are far to reject any mechanism differing from ours and simulating the cumulative effect. The only question is which of them is (or are) dominating mechanisms, and which give corrections only.
In this paper we make an attempt to show that the approach [7—12] may play the role of a leading mechanism.

II. The formulation of the space-time picture of cumulative-type processes. Basic relations

In order to realize any physical process, time and space intervals must be not zero. Consider the cumulation process from this point of view. We assume that as a result of the first act of inelastic interaction of an incident nucleon and the one of nucleus nucleons, a compound system is produced, and then decays into a given channel during small but finite time interval. So one should admit that there are possible subsequent inelastic collisions of this system with other nucleus nucleons increasing its mass and, hence, the greatest probable energy of a produced particle.

Thus, one may consider cumulation as a scheme describing "gathering"[3]) of nucleons by a primary particle into a compound-system with growing mass [7].

At first we consider the problem on the cross-sections of production of the compound-system including n nucleus nucleons and capable to emit a cumulative particle, i.e. the particle having a momentum much greater than the maximum one of NN-collision kinematics. Preliminary we express some ideas helpful for finding these cross-sections.

First of all, for production of a cumulative particle, a compound system is to have the maximal large mass. If the concept of the inelasticity coefficient, k, is used, the picture is to be close to that one when every act of an inelastic interaction occurs during the process of mass enlarging as $k \to 1$. However, it should be noted that such a process must not necessarily end in the formation of a total statistical system. The analogous situation takes place, as is known, in non-relativistic nuclear physics too.

The partial cross-section of such a process $\sigma(k \to 1) \simeq \sigma_c$ is to be

$$\sigma_c \ll \sigma_{NN}^{\text{in}}. \tag{1}$$

The attempts to estimate σ_c for pp-collisions were performed long ago [*16*]. They gave the value:

$$\sigma_c \approx (0.1 \div 0.2)\, \sigma_{NN}^{\text{in}} \tag{2}$$

within the interval:

$$5 \div 6\ \text{GeV} \leqq E_p \leqq 10 \div 12\ \text{GeV}. \tag{3}$$

The suggestion was made that the energy dependence of σ_c in the form

$$\sigma_c \sim \gamma_{\text{c.m.s.}}^{-2} \tag{4}$$

was possible.

These data were extremely inaccurate and referred to a narrow energy interval. Besides, very often the value of σ_c was associated with the production cross-section of the total statistical system [*16*]. Thus, magnitudes given above might be considered only as preliminary ones giving the order of magnitude σ_c. It is evident that σ_c is one of the determining characteristics of the model.

On the other hand, it is clear that the parameter σ_c cannot be the only one in the cumulation picture. In fact, the produced compound system may immediately decay with emitting energetic particle or its energy dissipates over many internal degrees of freedom producing further several "soft" particles. In both cases such a system will be "lost" for a cumulative process.

One may assume that the mean time $\bar{\tau}_0$ of emission of a cumulative particle, on the one hand, is to be less than the mean time of the decay of the thermodynamically equilibrium system $\bar{\tau}$ and, on the other hand, greater or equal to the time of a collision, $\bar{\tau}_{\text{col}}$ (the latter is only for the limiting case, as some time interval is needed for the rearrangement of a hadronic system)

$$\bar{\tau}_{\text{col}} \leqq \bar{\tau}_0 \leqq \bar{\tau}. \tag{5}$$

For $\bar{\tau}_{\text{col}}$ one may adopt

$$\bar{\tau}_{\text{col}} \approx \langle r_N \rangle / c \cdot \gamma_{\text{cms}}, \tag{6}$$

---

[3]) A. M. Baldin pointed out, practically simultaneously, the principal possibility of such a mechanism.

where $\langle r_N \rangle$ is the nucleon average radius. Initially [7] for $\bar{\tau}$ its estimation was taken following Pomeranchuk's model:

$$\bar{\tau} \approx r_0 \langle n_s \rangle^{1/3}/c, \tag{7}$$

where r_0 is the radius of nuclear forces, $\langle n_s \rangle$ is the multiplicity at $k \to 1$. Now based on the work [15] we may take for $\bar{\tau}$:

$$\bar{\tau} \approx (r_0 + l_h)/c \approx 8 \cdot 10^{-24} \text{ s} \tag{8}$$

where $l_h \simeq 1.2$ fm is the hadronization length.
We assume

$$\bar{\tau}_0 \approx \tau_0/\sqrt{s}. \tag{9}$$

In (9) we take into account the possibility of decreasing of the mean time of decay into a given channel when compound-system mass grows. The dependence (9) is prompted by the uncertainty relation and by the observed general tendency of varying lifetime of resonances with increasing their masses.
"Surviving" of states capable to emit cumulative particles with respect to the processes both of the total-system decay and of energy dissipation over the other degrees of freedom, may be defined by the usual exponential dependence:

$$\eta_{\text{cms}} \simeq \exp(-t/\bar{\tau}_0). \tag{10}$$

Passing to the laboratory reference frame and expressing t in terms of coordinates and speed, we obtain

$$\eta(z) = \exp\left[-\frac{z - z_{\text{in}}}{\bar{\tau}_0 (V\gamma)_{\text{cms}}}\right] \tag{11}$$

where z_{initial} is a coordinate of the system production or of the act of mass enlarging. Since in subsequent acts the product $\tau_0 V\gamma$ changes, one may put

$$\eta(z) = \exp\left[-\frac{z - z_{\text{in}}}{(\bar{\tau}_0 V\gamma)_n}\right] = \exp\left[-a_n(z - z_{\text{in}})\right] \tag{12}$$

(n is the number of "gathered" nucleus nucleons). For the same reason, in (9) one should use S_n instead of S. Then by kinematics we obtain:

$$a_n = \left[\frac{\tau_0 c}{\sqrt{s_{n'}}} \cdot \frac{(E_p{}^2 - m_p{}^2)^{1/2}}{\sqrt{S_{n'}}}\right]^{-1} = \frac{1}{\tau_0 c} \, \frac{(n'^2 + 1)\, m_p{}^2 + 2n' m E_p}{(E_p{}^2 - m_p{}^2)^{1/2}}, \quad n' = n - 1. \tag{13}$$

Now we may turn to the basic relations describing the process within the picture formulated. The scheme of the process is depicted in fig. 1 for a definite impact parameter, b, of a particle colliding with a nucleus (the case $p + A \to \pi + X$ has been considered). The probability of the compound-system production during the process of "gathering" of nucleons with a given distribution of points z_n where acts of its mass enlarging come about, may be given by the following expression:

$$W^{(n)}(b; z_1, \ldots, z_n) = w_1(z_1) \ldots w_n(z_n) \tag{14}$$

where $w_1(z_1), \ldots, w_n(z_n)$ are probabilities of acts at points $z_1, \ldots, z_n$, respectively. Coordinates of events are distributed over the interval $[z_{\min}, z_{\max}]$ under the condition:

$$z_{\min} = -\sqrt{R^2 - b^2} \le z_1 \le z_2 \ldots \le z_n \le z_{\max} = \sqrt{R^2 - b^2}, \tag{15}$$

R is the radius of nucleus with atomic weight A ($R = r_0A^{1/3}$, $z_0 = 1.2$ fm). The account of all possibilities of distribution of acts over point under the condition (15) leads to the integral.

$$W^{(n)}(b) = \int_{z_{\min}}^{z_{\max}} w_1(z_1)\, dz_1 \int_{z_1}^{z_{\max}} w_2(z_2)\, dz_2 \ldots \int_{z_{n-1}}^{z_{\max}} w_n(z_n)\, dz_n . \tag{16}$$

For the total cross-section of n nucleon "gathering" we have:

$$W^{(n)} = 2\pi \int_0^R bdb W^{(n)}(b) . \tag{17}$$

When calculating (16), we use simple approximations widely accepted in nuclear optics. First, we shall consider the nucleus as a hard sphere uniformly filled by matter with the density

$$\varrho = \begin{cases} A/V_A = 3\pi/4\pi r_0^3 , & r < R \\ 0, & r \geqq R . \end{cases} \tag{18}$$

Second, we consider the trajectories both of a primary particle and of compound systems produced by it, as rectilinear ones.
Third, following the analogy with optics, we take for $w_n(z)$

$$w_n'(z) = \sigma_c\varrho \exp\left[-\sigma_c\varrho(z - z_{n-1})\right] . \tag{19}$$

However, (19) does not take into account the possibility of the decay of a system (and, hence, its going away from the channel considered) between acts of the mass enlarging. Its effective account will be produced by multiplying (19) by time factors (12) for $n \geqq 2$. Then for $w_n(z)$ we have:

$$\begin{aligned} w_1(z_1) &= \sigma_c\varrho \exp\left[-\sigma_c\varrho(z_1 - z_{\min})\right], \\ w_{n+1}(z_n) &= \sigma_c\varrho \exp\left[-\sigma_c\varrho(z_n - z_{n-1})\right] \cdot \exp\left[-a_n(z_n - z_{n-1})\right] . \end{aligned} \tag{20}$$

Straightforward integration in (16) and (17) with $w_n(z_n)$ in the form (20) leads to the following expressions for $W^{(n)}$:

$$W_A^{(1)} = \frac{\pi}{2}\left\{2R^2 - \frac{1}{(\sigma\varrho)^2}\gamma[2; 2\sigma\varrho R]\right\},$$

$$W_A^{(2)} = \frac{\pi}{2}\frac{\sigma\varrho}{\sigma\varrho + a_2}\left\{2R^2 - \frac{1}{(\sigma\varrho)^2}\frac{\sigma\varrho + a_2}{a_2}\cdot\gamma[2; 2\sigma\varrho R] + \frac{\sigma\varrho}{a_2(\sigma\varrho + a_2)}\cdot\gamma[2; 2(\sigma\varrho + a_2)R]\right\}$$

$$W_A^{(3)} = \frac{\pi}{2}\frac{(\sigma\varrho)^2}{(\sigma\varrho + a_2)(\sigma\varrho + a_3)}\left\{2R^2 - \frac{k_1}{(\sigma\varrho)^2}\gamma(2; 2\sigma\varrho R] + \frac{k_2}{(\sigma\varrho + a_2)^2}\gamma[2; 2(\sigma\varrho + a_2)R] - \frac{k_3}{(\sigma\varrho + a_3)^2}\gamma[2; 2(\sigma\varrho + a_3)R]\right\}$$

where

$$k_1 = 1 + k_2 - k_3 ,$$

$$k_2 = \frac{\sigma\varrho(\sigma\varrho + a_3)}{a_2(a_3 - a_2)},$$

$$k_3 = \frac{\sigma\varrho(\sigma\varrho + a_2)}{a_3(a_3 - a_2)};$$

$$W_A^{(4)} = \frac{\pi}{2}\frac{(\sigma\varrho)^3}{(\sigma\varrho + a_2)(\sigma\varrho + a_3)(\sigma\varrho + a_4)}\left\{2R^2 - \frac{l_1}{(\sigma\varrho)^2}\cdot\gamma[2; 2\sigma\varrho R] + \frac{l_2}{(\sigma\varrho + a_2)^2}\cdot\gamma[2; 2(\sigma\varrho + a_2)R] - \frac{l_3}{(\sigma\varrho + a_3)^2}\cdot\gamma[2; 2(\sigma\varrho + a_3)R] + \frac{l_4}{(\sigma\varrho + a_4)^2}\cdot\gamma[2; 2(\sigma\varrho + a_4)R]\right\},$$

where

$$\begin{aligned} l_1 &= 1 + l_2 - l_3 + l_4, \\ l_2 &= \frac{\sigma\varrho(\sigma\varrho + a_3)(\sigma\varrho + a_4)}{a_2(a_3 - a_2)(a_4 - a_2)}, \\ l_3 &= \frac{\sigma\varrho(\sigma\varrho + a_4)(\sigma\varrho + a_2)}{a_3(a_3 - a_2)(a_4 - a_3)}, \\ l_4 &= \frac{\sigma\varrho(\sigma\varrho + a_2)(\sigma\varrho + a_3)}{a_4(a_4 - a_3)(a_4 - a_2)}. \end{aligned} \tag{21}$$

In formulas (21) $\gamma[n, \omega]$ is an incomplete gamma-function.
The expressions (21) are rather cumbersome but convenient for numerical calculations. The cross-sections $W_A^{(n)}$ (at least for $n \leqq 4$) can be written in a more compact way [7]:

$$W_A^{(n)} = \frac{\pi}{2}\frac{(\sigma\varrho)^n}{\prod\limits_{\varkappa=1,\dots,n}(\sigma\varrho + a_\varkappa)}\left\{2R^2 + \sum_{k=1}^{n} C_k^{(n)}\frac{\gamma[2; 2(\sigma\varrho + a_k)R]}{(\sigma\varrho + a_k)^2}\right\}, \tag{22}$$

and

$$a_1 \equiv 0,$$

$$C_{k+1}^{(n)} = \prod_{\substack{\lambda=1,\dots,n \\ \lambda \neq k}} (\sigma\varrho + a_\lambda)\left[\prod_{\substack{\zeta=1,\dots,n \\ \zeta \neq k}} (a_k - a_\zeta)\right]^{-1},$$

$$C_{k=1}^{(n)} = -\left[1 + (-1)^n \sum_{k=2}^{n} c_k^{(n)}\right].$$

So, formulas (21) and (22) solve the problem on the cross-section of the compound-system production within the process of n nuclear nucleons gathering.
The following important problem is how to describe the decay of such systems for the magnitudes of momenta of produced particles close to the kinematical limit. For this purpose, already the first studies of cumulative effects [17] used the assumption on scale invariance of the spectrum of produced particles. In the given case it means that the distribution over $x = p_\parallel/p_\parallel^{\max}$ is invariant with respect to the intermediate-system mass. In [7] such an assumption was also used.
However, now it is seen that the exact realization of scale invariance is not necessary in the model. It is enough to suggest the considerably weaker condition: the invariant inclusive cross-section in both hadron-hadron and collective-hadron interactions is approximately the same for equal masses of the intermediate compound system.
Further, in accordance with [12], the assumption on validity of this condition will be called the hypothesis on identity of hadronic inclusive spectra, or shortly, the identity

hypothesis. The model of cumulation processes based on the identity hypothesis posses- ses evident advantages.
Firstly, the validity of the model does not depend on the degree of fulfilment of scale invariance (especially it is essential at comparatively low energies where scale invariance does not take place often [*18*]).
Secondly, it makes closer the given model of cumulative process to the space-time formulation of multi-hadron production process [*13—15*].

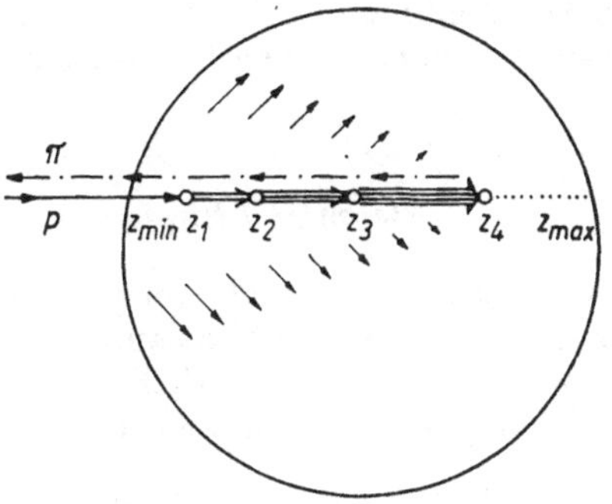

Fig. 1. Scheme of the process of "gathering"

Thus, when describing the decay of a compound-system, we shall use the invariant inclusive densities for production of particles of an i-th sort in pp- (or πp-) collisions:

$$\varrho_i(s, x_n, p_\perp) = \frac{E}{\sigma_{pp}^{in}} \frac{d^3\sigma}{dp^3}, \tag{23}$$

where

$$x_n = p_{\|,(i)}/p_{\|,(i)}^{max}(n) \tag{24}$$

and $p_{\|,(i)}^{max})n)$ — is the maximal magnitude of the longitudinal momentum of an i-particle for a given $p_\perp$. Its magnitude is defined by the kinematics of cumulative process of n-th order. At $p_\perp = 0$ (23) turns into the invariant cross-section of the particles flying out at angle $\theta = 0°$ and 180° with respect to the direction of motion of an incident particle. In the case of $\theta = 180°$, for instance, for $p + A \to \pi + X$ process for $p_{\|,(\pi)}^{max}(n)$ we have:

$$p_{\|,(\pi)}^{max}(n) \approx E_\pi^{max}(n) = n\,\frac{m_p}{2}\,\frac{\left(1 - \dfrac{m_p}{E_p}\right)\left(1 + \dfrac{m_p}{2nE_p}\right)}{1 + \dfrac{n^2+1}{2n} \cdot \dfrac{m_p}{E_p}} \tag{25}$$

m_p is the proton mass, E_p its total energy in the laboratory frame.
For the function $\varrho(S, x_n, p_\perp)$ sometimes the approximation is used which factorizes dependences on longitudinal and transverse momenta:

$$\varrho_i(S, x, p_\perp) = F_i(x, S) \cdot \exp(-a_i p_\perp{}^2). \tag{26}$$

In particular, for pions, $a_\pi \approx 4.24$ (see „fit 1" [*19*]).
One can determine approximately the pp-collision energy at which, according to the identity hypothesis, it is necessary to use the data on dependence of the invariant in-

clusive spectrum on the longitudinal momentum, that is on X. It has been calculated by equating masses of intermediate systems in pp-collision and in cumulative process:

$$s = 2m(\tilde{E} + m) = (n^2 + 1)\, m^2 + 2nmE_p{}^{\text{in}} \tag{27}$$

i.e.

$$\tilde{E} = \frac{1}{2m}\,[(n^2 + 1)\, m^2 + 2nmE_p{}^{\text{in}}) - m. \tag{28}$$

For example, at $E_p{}^{\text{in}} = 8.6$ GeV and $n = 2,3$, according to (28), we obtain $\tilde{E} \simeq 19 \div 30$ GeV. The data on production of various particles in this interval are contained, e.g., in [20].

If there is no dependence of F on s,

$$F(s, x) \simeq F(x), \tag{29}$$

— we obtain scale invariance. It seems (29) is valid for pion production with $x > 0.5$ [19]. Having considered the problems of formation of a compound system "absorbing" n nuclear nucleons and of description of its decay, we shall put down the invariant cross-section of production of cumulative particles on a nucleus in the form:

$$R_A{}^{(i)} = \sum_n \varrho_i(s, x_n, p_\perp) \cdot W_A{}^{(n)}. \tag{30}$$

The relation (30) coincides in form with the corresponding expression in [1]. However, in contrast to the latter, the identity hypothesis is used in (30) instead of scale invariance (which is its particular case). Besides, weight functions, $W_A{}^{(n)}$, being a consequence of the "gathering" scheme differ, in essence, from the corresponding expressions $P^{(n)}$ in [1] having the combinatorial nature what is caused by applying the fluctuation model in the paper mentioned.

III. Some general consequences of the model

Before applying the model formulated above for the analysis of experimental data, it is useful to dwell on some rather general conclusions following from its general relations and axiomatics.

1. A-dependence of invariant cross-sections

For the study of the dependence of the invariant cross-section on the mass number of a nucleus-target one should consider the behaviour of the functions $W_A{}^{(n)}$ in (30). We shall analyze two extreme cases: small and large time, i.e. $a_n > 1$ and $a_n \ll 1$ (σ_c is assumed to be small). So as $\sigma \ll \sigma_{NN}^{\text{in}}$, at least for, light and mediate nuclei:

$$\omega = 2\sigma\varrho R < 1. \tag{31}$$

Then for the case of small time ($a_n > 1$) from (21) and (22) it follows:

$$W_A{}^{(n)} \approx \frac{\pi}{2}\, \frac{(\sigma\varrho)^{n+1}}{\prod\limits_{\varkappa=1,\ldots,n} (\sigma\varrho + a_\varkappa)} \cdot \left\{2R^2 - \frac{1}{(\sigma\varrho)^2} \cdot \gamma[2;2\sigma\varrho R]\right\} \approx \frac{4\pi}{3}\, R^3\, \frac{(\sigma\varrho)^{n+1}}{\prod\limits_{\varkappa=1,\ldots,n} (\sigma\varrho + a_\varkappa)} \sim A. \tag{32}$$

(The explicit expression for the function $\gamma[2, \omega]$ and its decomposition in a small, according to (31) parameter ω are used). The expression (32) indicates that in any n-th order of cumulation the dependence of the invariant cross-section is proportional to the nuclear volume. In the widely accepted approximation of A-dependence by the function:

$$R \sim A^N \tag{33}$$

it implies that the power index equals

$$N \approx 1. \tag{34}$$

In the case of large time $a_n \ll 1$ it is more convenient to perform the integration in (16) assuming $\eta \equiv 1$. Then

$$\begin{aligned}
W_A^{(1)} &= \frac{\pi}{2(\sigma\varrho)^2}\left\{\frac{1}{2}\,\omega^2 - \gamma[2,\omega]\right\},\\
W_A^{(2)} &= \frac{\pi}{2(\sigma\varrho)^2}\left\{\frac{1}{2}\,\omega^2 - \gamma[2,\omega] - \gamma[3,\omega]\right\},\\
W_A^{(3)} &= \frac{\pi}{2(\sigma\varrho)^2}\left\{\frac{1}{2}\,\omega^2 - \gamma[2,\omega] - \gamma[3,\omega] - \frac{1}{2}\,\gamma[4,\omega]\right\},\\
W_A^{(4)} &= \frac{\pi}{2(\sigma\varrho)^2}\left\{\frac{1}{2}\,\omega^2 - \gamma[2,\omega] - \gamma[3,\omega] - \frac{1}{2}\,\gamma[4,\omega] - \frac{1}{6}\,\gamma[5,\omega]\right\}.
\end{aligned} \tag{35}$$

Supposing, as before, that (31) is valid and decomposing incomplete gamma-functions $\gamma[n, \omega]$ into the series

$$\gamma[n,\omega] = \sum_{\lambda=0}^{\infty} \frac{(-1)^\lambda\,\omega^{\lambda+n}}{\lambda!(\lambda+n)}; \tag{36}$$

and retaining the two leading terms, we obtain

$$\begin{aligned}
W_A^{(1)} &\approx \frac{4\pi}{3}\,R^3(\sigma\varrho)\left[1 - \frac{3}{4}\,\sigma\varrho R\right] \sim A,\\
W_A^{(2)} &\approx \pi R^4(\sigma\varrho)^2\cdot\left[1 - \frac{16}{15}\,\sigma\varrho R\right] \sim A^{4/3},\\
W_A^{(3)} &\approx \frac{8}{15}\,\pi R^5(\sigma\varrho)^3\left[1 - \frac{5}{4}\,\sigma\varrho R\right] \sim A^{5/3},\\
W_A^{(4)} &\approx \frac{2}{9}\,\pi R^6(\sigma\varrho)^4\cdot\left[1 - \frac{11}{9}\,\sigma\varrho R\right] \sim A^2.
\end{aligned} \tag{37}$$

Thus, in this case

$$W_A^{(n)} \sim (\sigma\varrho)^n\,R^2\cdot R^n \sim A^{2/3+n/3} \tag{38}$$

and the power index N equals

$$N = (2+n)/3. \tag{39}$$

The dependences obtained are given in fig. 2. The "volume" dependence (32), (34) is explained by the fact that at small times "gathering" comes about in a relatively small region of a nucleus and, by an order of magnitude, $\overline{\Delta z} \approx 1/a_2 \approx$ const. (R). In this case the process is "local". At large times $W_A^{(1)} \sim A$, as σ is small. Further increase of N with increasing n by 1/3 (see (39)) is caused by the fact that the contribution of every order

of cumulation is proportional, roughly speaking, to the effective free path in a nucleus $R \sim A^{1/3}$.
Thus, at a fixed E^n the increase of time interval $\bar{\tau}_0$ (i.e. the increase of the admissible extent of energy dissipation in a compound system) for one cumulative process with respect to another, should strengthen A-dependence of the invariant cross-section, i.e. N:

$$N = f(\bar{\tau}_0) \qquad \text{growing function.} \tag{40}$$

Certainly, the actual A-dependence is to be determined by the model parameters σ and $\bar{\tau}_0$ calculated by comparing with data on cumulative particles spectra. It is apparent that it is to be described by the N-behaviour depicted qualitatively in fig. 2 by the dashed curve which is intermediate between two extreme cases considered.

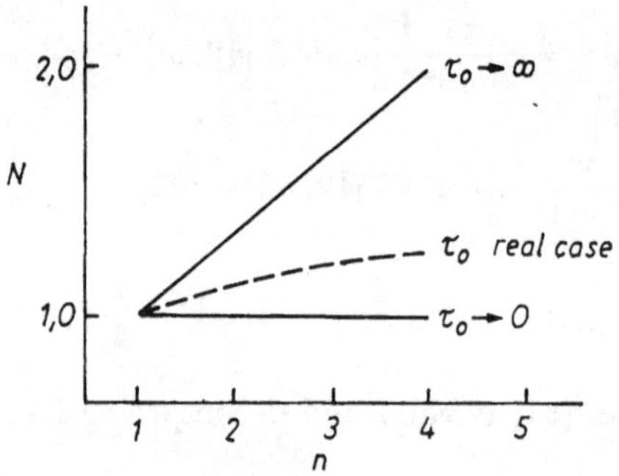

Fig. 2. A-dependence in extreme cases

2. On energy dependence of the invariant cross-section

Let us consider the energy behaviour of $R^{(i)}(\theta = 180°)$ under the following assumptions. First, the incident energy is high enough, i.e.

$$E^{\text{in}} \rangle\rangle\rangle\, m_p = m. \tag{41}$$

Second, the cross-section of production of a compound-system, σ, is as before, small and does not depend on energy:

$$\sigma \approx \text{const.}\,(E^{\text{in}}). \tag{42}$$

Third, the approximate scale invariance of inclusive spectra holds.
Using (41) we have from (13) for a_n:

$$(a_n)_{E^{\text{in}}\rangle\rangle\rangle m} \approx n' \frac{2m}{\tau_0 c} = (n-1)\frac{2m}{\tau_0 c} = \text{const.}\,(E^{\text{in}}). \tag{44}$$

From (42) and (44) it follows that

$$W_A^{(n)} = \text{const.}\,(E^{\text{in}}) \tag{45}$$

On the other hand, x_n also seems to be independent of energy. It is not difficult to observe, using (25) in the limit (41):

$$p^{\max}_{\|,(\pi)}(n) = n\,\frac{m}{2} \simeq \text{const.}\,(E^{\text{in}}). \tag{46}$$

Then from (43), (45) and (46) we obtain the asymptotic behaviour:

$$[R_A{}^{(i)}]_{n=\text{const.}} = \text{const.}\ (E^{\text{in}}). \tag{47}$$

These conclusions are valid at comparatively low energies for light particles, pions.

3. The $R(P^{\text{in}})$-dependence as a function of particle mass

Now let us clear up the question about the behaviour of $R_A(P^{\text{in}})$ as a function of the mass of a produced cumulative particle. For this purpose it is convenient to consider the dependence of $R_{Pb,n}(T = 0.2\ \text{GeV}, \theta = 180°, P^{\text{in}}) = F_{\pi;\bar{p}}(x_n)\ W_{Pb}^{(n)}$ for two cases: pions (light particles) and antiprotons (heavy particles) production.
We shall assume, as before, that $W^{(n)} \approx \text{const.}\ (P^{\text{in}})$ (see (45)).

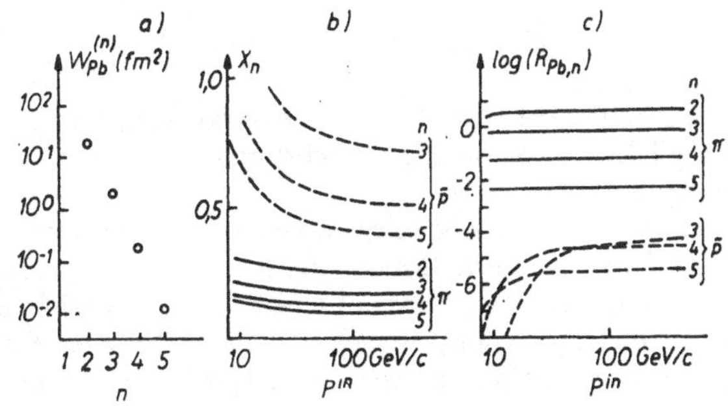

Fig. 3. Dependence of $R(\theta = 180°, T = \text{const.}, P^{\text{in}})$ for particles with different masses

In fig. 3a the circles represent $W_{Pb}^{(n)}$ calculated at $\sigma_c = 0.25\ \sigma_{pp}^{\text{in}}$. In fig. 3b the magnitudes of $x_n{}^{(\pi;\bar{p})}(P^{\text{in}})$ are given, and in fig. 3c $R_{Pb,n}(P^{\text{in}})$[4]) obtained from data of fig. 3a, b and $F_{\pi,\bar{p}}(x)$ falling down sharply with growing x built on the basis of the information available in [20] are represented.
We see that, in contrast to the case of pions, functions $R_{Pb,n}$ for antiprotons come to the asymptotic regime at much larger P^{in} ($P^{\text{in}} \simeq 70 \div 100$ GeV/c). The smallness of yield of antiprotons emitted at small P^{in} is caused (as seen in fig. 3a, b) both by large x_n and small cross-sections $W^{(n)}$ giving the effective contribution at these values of P^{in}. Hence, it follows that the search of cumulative particles, constituents of a heavy pair is easier to make at higher energies. More accurate estimations of R for these cases will be given in Sect. V.

4. Dependence of the onset of the asymptotic regime on mass number A

For primary particle energies at which inequality (41) is not strong enough (E^{in} exceeds m by a factor of $2 \div 4$) asymptotics (46) takes no place. Under these conditions accessory effects, neglected in the model, being an idealization may appear. These effects may manifest themselves in the character of going to asymptotics.
For example, in the model the energy losses of a primary particle in a nucleus, before it initiates cumulative process, are not taken into account. From the qualitative point of view it is quite clear that these losses increase with A (probably, proportionally to the

[4]) Hereafter the invariant cross-section R is expressed in mb · GeV^{-2} · c^3.

average free path, i.e. $\sim A^{1/3}$). So the onset of the asymptotics with respect to energy, E^{in}, is to be observed at different A beginning from various E^{in}. Hence, it follows that the onset of asymptotics of the cumulative process is earlier for light nuclei than for heavy ones.

As is seen from § 3. this conclusion is valid when produced particles are pions. For antiprotons the asymptotics onset occurs at $E^{in} \approx 70 \div 100$ GeV/c (see fig. 3) when the energy losses for excitation of a nucleus can be neglected.

5. On dependence of the form of a cumulative particles spectrum on mass number A

The experimental data indicate that the spectrum of cumulative particles has approximately the exponential form. For instance, the empirical description of the cumulative-pions spectrum at an angle about $\theta \approx 180°$ often uses the approximation

$$R_A(T_\pi) \sim \exp(-T_\pi/T_0). \tag{49}$$

The theoretical spectrum also is to be close to the exponential one. Indeed, for $x > 0.5$ the invariant density ϱ falls off exponentially with growing x. $W^{(n)}$ as functions of n also decrease approximately exponentially.

On the other hand, according to § 1. one may expect a smooth growth of the power index N (see fig. 2, the dashed line) with n. Hence, the increase of cumulative particles yield with growth of A is more intensive for those of them flying out with large T (as for them N is larger). In terms of approximation (49) it means that at fixed energy E^{in} the parameter T_0 is, generally speaking, an increasing (probably, weakly increasing) function of A:

$$T_0 = f(A). \tag{50}$$

6. On angular distribution of cumulative particles

The problem on angular distribution of cumulative particles is of the greatest interest. Just here they may be detected quite definitely. And the most interesting physical factors may manifest themselves.

When calculating angular distributions one may use the invariant density written in the factorized form (26).

However, both in this case and in the estimation of the absolute magnitude of the yield the problem of cumulative particles produced in a nucleus passing through it is rather important. Unfortunately, in overwhelming majority of models this problem even is not mentioned. Let us take pions production as an example and consider what is the essence of this problem and how to solve it. Let us turn to fig. 1. A pion produced, say, in z_4 is to cover a noticeable, on the average, path $z_4 - z_{min}$. The question arizes whether pion can fly out of a nucleus, practically, without absorption or the absorption should be taken into account. There exists the possibility to neglect this effect for energetic pions [*10*]. So, as the mean free path of a primary proton $\Lambda \approx 1/\varrho\sigma_{NN}^{in} \approx 2$ fm $\approx r_0$, it interacts, practically, with each nucleon along the trajectory of its motion. Nucleons acquiring, on the average, the transverse momentum $\langle p_\perp \rangle \approx 0.4$ GeV/c are to be displaced to the side of the trajectory of the proton (arrows in fig. 1) making free the channel in the nucleus. One can roughly estimate from below the displacement of l nucleons. The time for "gathering" from piont z_1 to z_4 and for covering this interval ($z_4 \to z_1$) by a relativistic meson is

$$t \approx 2\overline{\Delta z}/c. \tag{51}$$

Hence the displacement:

$$l \approx t \cdot v_{N,\perp} \approx \frac{2\overline{\Delta z}\langle p_\perp \rangle_N}{mc} \simeq 2.5\ \text{fm} \approx r_0 \tag{52}$$

(we assume $\overline{\Delta z} \approx 3$ fm).
Thus, the density of matter in the channel falls sharply, and the processes of "absorption" turn to be unimportant. (Thus effect of "trailing" was used earlier in attempts of the description of the multiple production process developing into the forward hemisphere [21] at considerably high energies. However, just in this case it does not "work". l is negligibly small due to the small difference in paths of particles of "leading" and "pionization" groups). So, as the first, rough approximation, the effects of absorption can be neglected.
Nevertheless, pions passing in the channel surrounded by the dense absorptive medium may change their angular distribution at the expense of other diffraction effects.
Parameters of the diffraction picture are determined by the wavelength of a pion, by the distance between an emitter (a compound-system) and yield off the channel, and by the number of Fresnel's zones in yield hole, i.e. by the pion momentum, by the length of gathering range, and by the diameter of the yield hole. One can show that one of the angles at which the destructive influence of two first Fresnel's zones may be revealed is determined by the expression:

$$\theta \approx \pi - \arctan\left[\frac{\lambda_\pi}{\Delta z}\right]^{1/2} \simeq \pi - \arctan\left(\frac{h}{p_\pi \Delta z}\right)^{1/2}. \tag{53}$$

For cumulative pions with $p_\pi > 0.4$ GeV/c and at $\Delta z \approx 3$ fm we have

$$\theta \approx 150^\circ \div 170^\circ. \tag{54}$$

As the quantities determining θ may fluctuate noticeably, it should hardly be expected that the diffractive picture will be clear. Its clearness may depend on energy and sort of particles and may become better at definite combination. Thus, if one has no extensive experimental and theoretical studies of this problem, one can only state at present that the modulation of the angular distribution into a backward hemisphere at the expense of diffraction effects is possible.

7. Does the model contradict the experiment on production of $\mu^+\mu^-$-pairs on nuclei?

Paper [22] contains a lot of criticizm of cumulative models. For this purpose the data on reaction [23]

$$\text{p} + U \to \mu^+\mu^- + X, \qquad E_p^{\text{in}} = 28.5\ \text{GeV} \tag{55}$$

are used. Detailed measurements on the effective mass show that the kinematical limit for it, in the case of the process (55), does not differ practically from that for analogous process in pN-collisions at the same energy.
Based in this fact, authors of [22] make a conclusion that the cumulative models, as the flucton one (among them the model of a coherent tube and the hydrodynamical one) contradict the experiment. The presence of objects with double, triple, and so on mass of a nucleon in a nucleus should lead to sharp increase of the kinematical limit.
Let us consider this argument in view of our model.
First of all, it should be emphasized once more that the "gathering" model does not require the presence of prepared objects with masses exceeding the mass of a nucleon in a nucleus.

Keeping to the scheme adopted for $\mu^+\mu^-$-pairs production via an intermediate gamma-quantum and also to logics of the "gathering" model, one can assert that $\mu^+\mu^-$-pairs with maximal masses may be produced only in the first act on pN-collision. In the subsequent acts there occurs the collision between a heavy-compound-system with defrosted internal hadronic degrees of freedom, and a nucleon. The deceleration of the system is less effective. Hence, from kinematics it follows that the maximal possible energy of γ-quantum in this case is significently less than in the case of pN-collision.
In this consideration it was assumed that the whole mass of a compound-system did not convert into a photon. This assumption is justified to a great extent. Indeed, the return of an excited system back into the state only with two degrees of freedom, nucleons, is a fluctuation with the probability of an order of

$$w \approx \exp\left[-\left(\sqrt{s_{n=1}} - 2m_N\right)/T\right] \tag{56}$$

(see [*16*] in connection with the discussion of the statistical mechanism of elastic scattering). The probability of transformations of this energy into the energy of a hard photon is not greater than the probability (56). The estimation (56) gives at $E_p{}^{\text{in}} = 29.5$ GeV and $T \approx 2\mu_\pi$ [*16*]:

$$w_\gamma \leqq w_{2h} \simeq 10^{-8}, \tag{57}$$

a negligibly small value.
Thus, according to the "gathering" model the kinematical limit for the effective mass of $\mu^+\mu^-$-pair in a reaction of the type $\mathrm{p} + A \to \mu^+\mu^- + X$ must coincide with its magnitude in the case of reaction $\mathrm{p} + N \to \mu^+\mu^- + X$ at equal initial energies (of course, in a simple scheme of $\mu^+\mu^-$-pair production). As has already been pointed out this conclusion agrees with experiment. Hence it follows that the limiting characteristics of processes of the bremsstrahlung type in nuclei should coincide with those for nucleon-nucleon collisions.

IV. Comparison with experimental data

Omitting details (one can find these in the works cited) we shall concern only with the most substantial aspects of this problem.

1. The process $\mathrm{p} + A \to \pi + X$ and the parameters of the model

As it has been marked above, the model results depend on the choice of two parameters, σ and τ_0, An approximate order of their magnitudes is known beforhand (see Sect. II). Their more accurate values are to be established by comparing calculations with experiment. The comparison with data [*1*] on cumulative pions in the process:

$$\mathrm{p} + A \to \pi(\theta = 180^\circ) + X, \qquad E_p{}^{\text{in}} \approx 6 \div 8\ \text{GeV} \tag{58}$$

gives the values

$$\sigma \simeq 0.25\sigma_{NN}^{\text{in}}; \qquad \tau_0 \simeq 5\ \text{GeV} \cdot \text{s} \cdot c. \tag{59}$$

The comparison is shown in fig. 4. It should be noted at once that values (59), as is expected for the given energies are within the intervals (1), (5). From this result it follows also that the "gathering" length is

$$\overline{\Delta z} \simeq 1/a_2 \approx 2.5\ \text{fm} \tag{60}$$

i.e. much smaller than the radius of heavy nuclei and much larger than the core radius in nucleon-nucleon interaction.

In fig. 5 the theoretical [7] and experimental [1] data on A-dependence of cumulative pions, $R_A^{(\pi)}$, are compared at different values of variable $\varkappa = T/T^{\max}(NN \to \pi)$. As (60) takes place, then, as it should be expected for this case from the general consideration (see III, 1), it is close to the case $N \approx 1$.

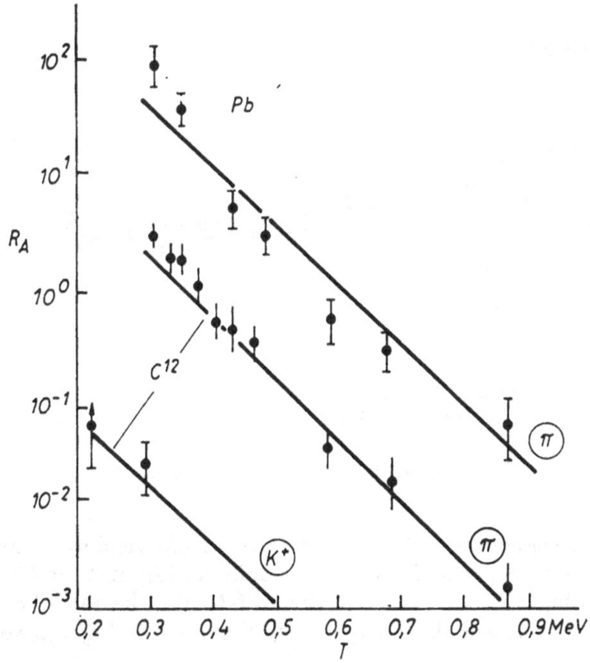

Fig. 4. Invariant cross section $R_A^{(\pi,K^+)}(\theta = 180°, T)$

In fig. 6 angular distributions of cumulative pions ($p_\pi = 0.5$ GeV/c [24]) in the back hemisphere are compared for process $\mathrm{p} + \mathrm{Pb} \to \pi + X$, $E_p^{\text{in}} = 8.6$ GeV. The calculation is made for the $p_\perp$-dependence of the invariant density represented by "fit 2" in [19]. We remind that the irregularity observed near $\theta \approx 170°$ may be associated with diffraction phenomena (see sect. III, 6).

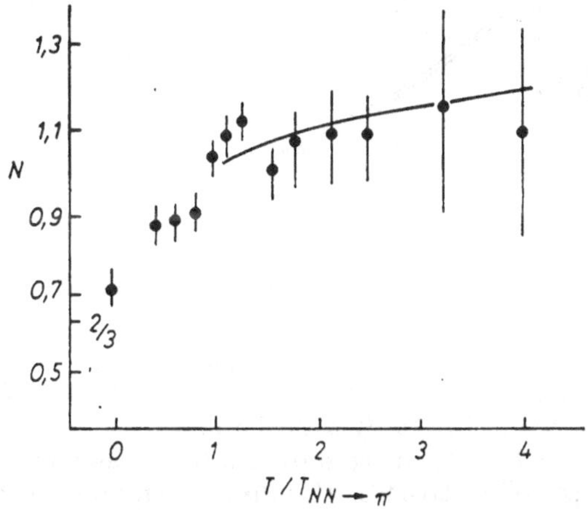

Fig. 5. Dependence $N(\varkappa = T/T^{\max}_{NN \to \pi})$

Generally speaking, both the cross-section σ and time $\bar{\tau}_0$ depend on energy. The dependence of the latter is postulated by (9) (see also fig. 7b). As for the dependence $\sigma(E^{in})$, there is nothing known. It is of great interest to establish it by means of the given model.

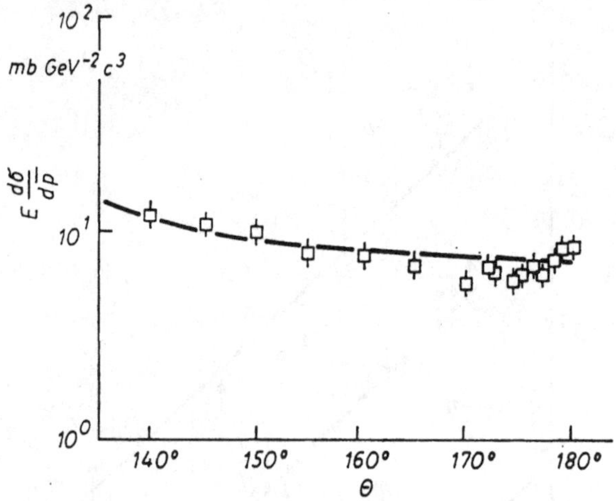

Fig. 6: $R_{Pb}(\theta_\pi, P_\pi = 0{,}5$ GeV/c, $P^{in} = 8.6$ GeV/c)

For this purpose, the calculations [12] of cumulative pions yield were carried out both for $P^{in} < 6 \div 8$ GeV/c (appropriate data are available in [25]) and for $P^{in} > 6 \div 8$ GeV/c. For instance, choosing σ so that at every $P^{in} < 6$ GeV/c the ratio $[R_A^{(\pi)}(P^{in})/R_A^{(\pi)}(P^{in} = 8 \text{ GeV}/c)]_{\varkappa=\text{const.}}$ coincide with its experimental value (see Fig. 8), we obtain function

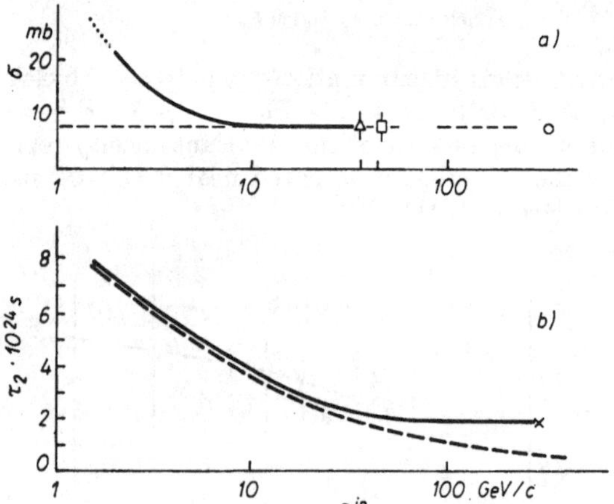

Fig. 7. Dependence of the cross-section of formation of the compound system a) and its life-time b) on energy E^{in}

$\sigma(P^{in})$. In principle, such a procedure could lead to different dependences of $\sigma(p^{in})$ for various $\varkappa$. However, the calculation carried out for $\varkappa = 1.0; 1.5; 2.0$ gave practically the same dependence depicted in fig. 7a by the solid line.

The form of dependence $R_A(P^{in})$ in fig. 8 directs to the assumption that in the range $P^{in} > 6$ GeV/c $\sigma \approx$ const. (P^{in}). To check this the data on cumulative mesons yield based on the measurements [5] and [6], were used. In [5] the process p + Ta $\rightarrow \pi + X$ has been studied at $P_{in} = 28.5$ GeV/c and the integral function $R_{Ta}(\theta = 180°) = \int_T^{T^{max}} E\left(\frac{d^3\sigma}{dp^3}\right) \times dT$ has been obtained. The calculation of this function at several values of the cross-section $\sigma = 10$, 8 and 6 mb and its comparison with experiment are displayed in fig. 9

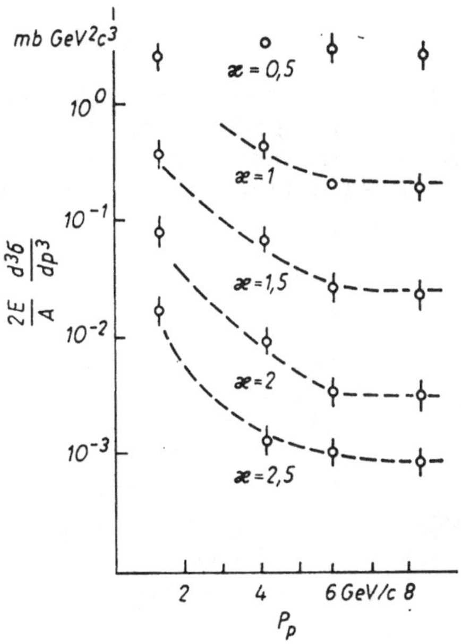

Fig. 8. Dependence $(1/A)\, R_A^{(\pi)}(\theta = 180°, \varkappa, P^{in})$

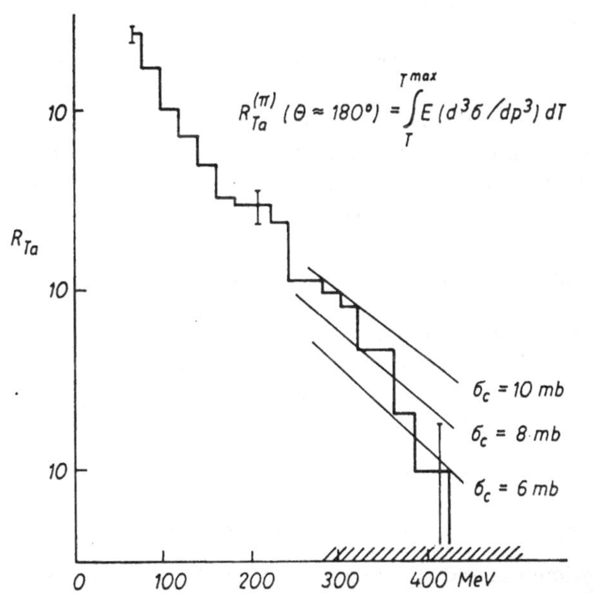

Fig. 9. $R_{Ta}(T)$ for the process $p + Ta \to \pi(\theta = 180°) + X$, $P^{in} = 28{,}5$ GeV/c

(the shaded region near the abscissa axis means the cumulative region). From this it follows that $\sigma \approx (8 \pm 2)$ mb; in fig. 7a this result is marked by the triangle.
The analogous treatment for the invariant cross-section of process $\pi + C^2 \to \pi^-(\theta > 90°) + X$ at $p_\pi^{in} = 40$ GeV/c on the basis of data [6] leads to the same results (the square in fig. 7a). Therefore, the data analysis justifies the assumption that $\sigma \simeq$ const. (P^{in}) in a wide energy interval.

2. Other cumulative particles

The information obtained on the model parameters behaviour can be used for the description of processes of production of particles other then pions.
In fig. 4 the comparison of experimental and theoretical data is made for process $p + C^2 + K^+(\theta = 180°) + X$. Figures 10 and 11 demonstrate the situation for processes

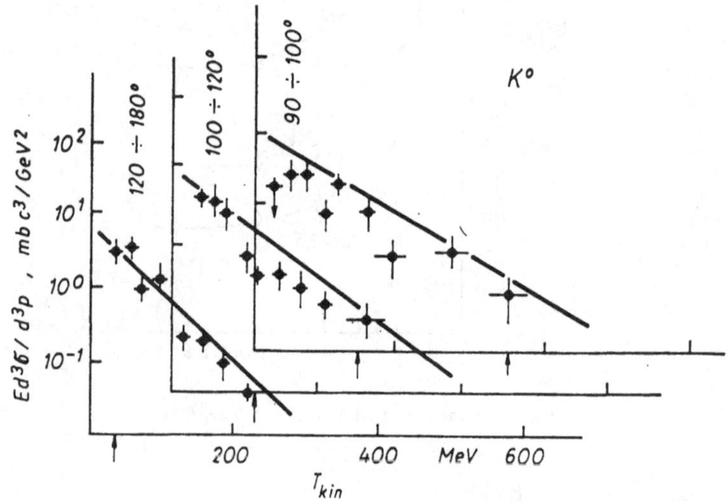

Fig. 10. $R(\theta, T)$ for the process π + (propane + xenon) $\to \Lambda° + X$; $P^{in} = 2.9$ GeV/c

π + (propane + xenon) $\to K° + X$ [26] and π + (propane + xenon) $\to \Lambda_0 + X$ [27] at $P_\pi^{in} = 2.9$ GeV/c. For σ there was taken the values corresponding to $P^{in} = 2.9$ GeV/c in fig. 7a multiplied by 0.7 (ratio $\sigma_{\pi N}^{in}/\sigma_{NN}^{in}$). Figures 4, 10 and 11 indicate the satisfactory description of spectral-angular characteristics of these cumulative particles.

3. The "gathering" model and the production of particles with large $p_\perp$

In [12] it was assumed that the model stated can be applied also to interpret the experiments [3, 4] on production of particles with large $p_\perp$ on nuclei at high energies. The analysis used the information on function $F^{(0)}(s, x_\perp)$, $x_\perp = p_\perp/2\sqrt{s}$ available in [4].
A satisfactory agreement is achieved at the parameters:

$$\sigma_1 = 8 \text{ mbarns}, \qquad \tau_0 = 15 \text{ GeV}/c \text{ s}$$

or (61)

$$\sigma_2 = 10 \text{ mbarns}, \qquad \tau_0 = 15 \text{ GeV}/c \text{ s}.$$

In fig. 12 the experimental and theoretical data for the invariant inclusive cross-sections of production of pions by protons are represented for $E^{in} = 300$ GeV on nuclei Be, Ti and W. Here additional comments are needed.

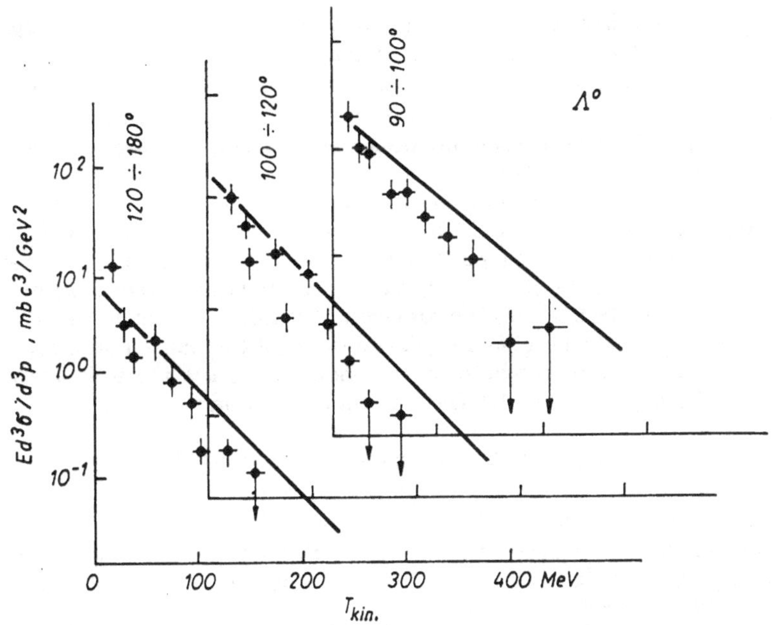

Fig. 11. $R(\theta, T)$ for the process π + (propane + xenon) → $\Lambda^\circ + X$; P^{in} = 2,9 GeV/c

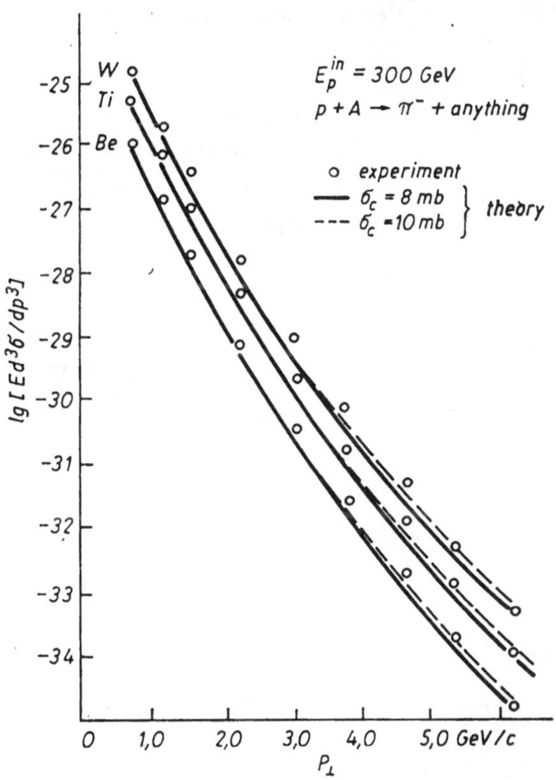

Fig. 12. Functions $R_A(p_\perp)$ for π-meson, E_p^{in} = 300 GeV

a) From (61) it follows that at $E^{\text{in}} = 300, 400$ GeV, $\sigma = 8 \div 10$ mb also, i.e. again there is no contradiction with the tendency depicted in fig. 7a (for the given case — the circle).
b) However, (61) points to the deviation from the dependence (9) for the surviving time of a compound-system. On the one hand, this may be associated with the fact that the model is not completely valid for the production of particles with large $p_\perp$. On the other hand, this may result from a quite natural reason. Indeed, the dependence (9) implies the decrease of time $\bar{\tau}_0$ with growing compound-system mass. At relatively low energies E^{in} it is acceptable. However, in the limit of extremely high energies, according to (9), $\bar{\tau}_0$ tends to zero. From the physical point of view, this limit is not realistic, as the surviving time is not less than the time of rearrangement of a compound-system before emitting of a particle which must be not less than some minimal value. The latter being multiplied by the light velocity, c, should give a size of an order of the hadron size. Indeed, from fig. 7b (the dotted line corresponds to the dependence (9), and the solid line stands for the result (61) at energies $300 \div 400$ GeV) we have $t_{\min}$ and:

$$\langle r \rangle \simeq t_{\min} \cdot c \simeq 2 \cdot 10^{-24}\,\text{s} \cdot 3 \cdot 10^{23}\,\text{fm/s} \approx 0.6\,\text{fm} \tag{62}$$

i.e. $\langle r \rangle$ coincides, in scale, with the hadron size and, hence, with sizes of a compound-system formed in overlapping their volumes (for more details on the calculation of this reaction see in [*12*]). From fig. 12 it follows also that the theory reproduces also the A-dependence of yield at large $p_\perp (> 3 \div 4$ GeV/c).
To all appearance, the results of experimental study [*28*] of the A-dependence for "jets" with large $p_\perp$ at $P^{\text{in}} = 200$ GeV/c are in close connection with the facts presented here. It is shown in fig. 13. The power index N for individual pions with large $p_\perp$ is rather lesser:

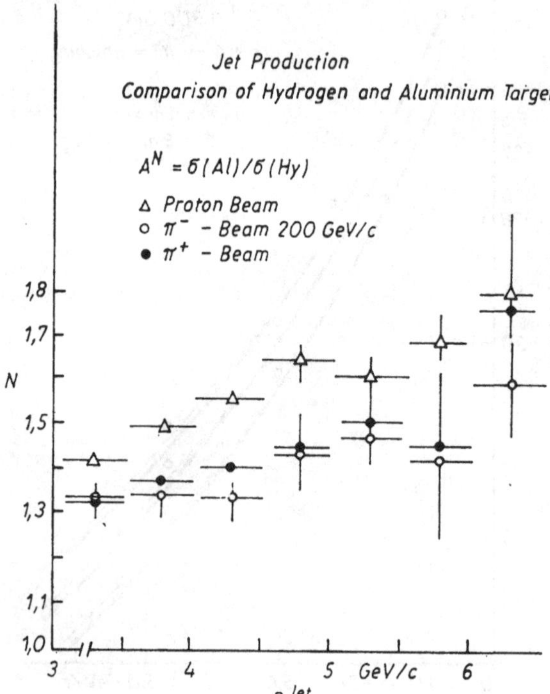

Fig. 13. Behaviour of N for "jets" with large $p_\perp$

in the range of $p_\perp > 3 \div 4$ GeV/c

$$N_\pi \approx 1.1 < N_{\text{jet}} \approx 1.3 \div 1.5 .$$

In principle, the model agrees with this observation if one suggests that the energy dissipation over larger number of degrees of freedom (and, thus, over larger particles number) requires still larger time. It means that time τ_0 for which a compound-system is still capable to emit "jets" may become greater than in the case of single pion emission. However, then the A-dependence for "jets", in fact, is to be more sharp (see Sect. III, § 1). Of course, the possibility of other interpretations cannot be excluded.

4. Cumulative processes involving deuteron

The case with deuteron requires a special analysis, as it is (even in a rough approximation) not an object with the uniform density and quite definite boundary. So, in the given case it is difficult to use directly the method stated above for pA- interaction. However, keeping to the spirit of the "gathering" model one can make here reasonable evaluations, as well.

Let us briefly discuss two types of experiments on the meson production in the reaction involving deuteron

$$\mathrm{p} + \mathrm{d} \to \pi^-(\theta = 180°) + X, \tag{63}$$

$$\mathrm{d} + A \to \pi^-(\theta = 0°) = X. \tag{64}$$

For both the cases the order of cumulation, $n = 2$.

Considering, for instance, the process (63), we have for the invariant cross-section:

$$R_d = 2E_\pi \frac{d^3\sigma}{dp^3} = \sigma \cdot 2F(x), \tag{65}$$

where σ is the cross-section of production of a compound-system by an incident proton. It can be approximately written down as

$$\sigma = \sigma_{p,d}^{\text{in}} \cdot W, \tag{66}$$

where $\sigma_{p,d}^{\text{in}}$ is the cross-section of inelastic (p,d)-interaction, and W is the probability of realization of the mechanism of gathering of three nucleons into a unique system. Then the invariant density equals:

$$\varrho_d = \frac{1}{\sigma_{p,d}} R_d = W \cdot 2F(x). \tag{67}$$

It is evident that W is determined by the behaviour of the distribution of nucleons in deuteron over relative distances, i.e. by $|\psi_d|^2$, the factor for going of the system out of the considered channel between "gathering" acts, and also by the orientation of the deuteron axis with respect to the axies of p,d-collision.

As $\psi_d = U_d(r)/r\sqrt{4\pi}$ it is convinient to use the wave function of deuteron in s-state with the hard core of radius r_c which is suggested in [29] which excludes the contribution of very small distances.

In the cylindrical reference frame (fig. 14) natural for the given geometry the expression for W may be written in the following way:

$$W(b^{\max}, a) = 2(2\pi) \int_0^{b^{\max}} b\, db \int_{z_{\min}}^{\infty} |\psi_d(b, z)|^2, e^{-az}\, dz, \tag{68}$$

where b is the impact parameter, exp $(-az)$-, a factor representing the time of dying out of states capable to emit a cumulative pion. The coefficients "2" takes into

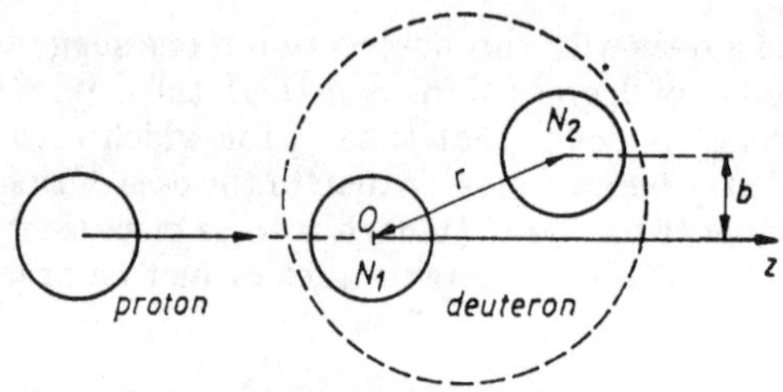

Fig. 14. Scheme of nucleon gathering in p,d-process

account the independence of the result of the nucleon index and also of the choice of $Z_{\min}$ in the internal integral.
The more detailed form:

$$W(b^{\max}, a) = 4\pi \int_0^{b^{\max}} b\, db \int_{z_{\min}}^{\infty} \frac{u^2(\sqrt{b^2+z^2})}{b^2+z^2}\, e^{-az}\, dz \tag{69}$$

$$Z_{\min} = \begin{cases} \sqrt{r_r^2 - b^2} = \sqrt{\Delta}, & \text{if } \Delta > 0 \\ 0, & \text{if } \Delta \leqq 0. \end{cases} \tag{70}$$

As the function $U(z)$ has a rather cumbersome form, we do not display it here (see [29]). According to (68) and (69) the function ϱ_d, as in the case of p,A-interaction, depends only on two parameters, a and $b^{\max}$. It is natural to choose them as in p,A-process. Then

$$a = a_2 \simeq 0.42 \text{ fm}^{-1} \tag{71}$$

$$b^{\max} \simeq 0.8 \text{ fm}$$

(for more details see [8]).
Then calculations lead to the value:

$$W(b^{\max} \approx 0.8;\quad a = 0.42) \approx 0.05. \tag{72}$$

In other words, the probability that in such a collision deuteron behaves kinematically as a unique object is $\approx 5\%$.
For invariant density we have respectively

$$\varrho_d \approx 0.1 \cdot F(x). \tag{73}$$

In fig. 15 the experimental data on ϱ_d-function at $P_P^{\text{in}} = 8.4$ and 6.0 GeV/c [25] and the theoretical curve corresponding to ϱ_d from (73) are represented (x has been calculated in the framework of kinematics oa p,d-collision).
Because of a rather approximate character of calculations one may think that the agreement of the data in fig. 15 is good.
Now a few words about the process (64) [19]. In contrast to the process (63) there was measured the probability of production of mesons flying out at forward angles and having the energy much exceeding the energy per one of deuteron nucleons. It has been established that the ratio of the cross-section of production of pions by deuterons to that one of production by protons at the same energy release depends neither on x, nor

on E_d and equals

$$\alpha \approx 0.06 \pm 0.007. \tag{74}$$

From the point of view of the picture used in p,d-process (63), this result admits a simple interpretation. Indeed, the fraction of the cases (72), when deuteron behaver kinematically as a unique object practically equals the value calculated before and having the same sense.

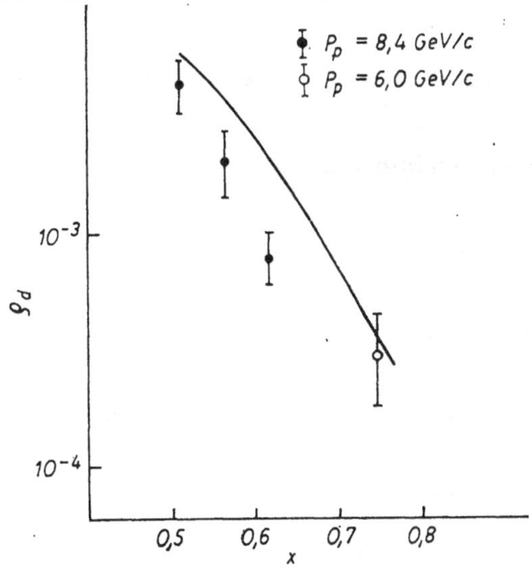

Fig. 15. Invariant density $\varrho_d(x)$

5. Sub-threshold production of antiprotons

A possible manifestation of effects of collectivization of hadron interaction in the nuclear matter is the production of antiprotons in collisions of relativistic protons with a nucleus at sub-threshold energies.

The threshold energy of antiproton production in nucleon-nucleon collision equals:

$$E_p^{\text{thr.}} = 6.58 \text{ GeV}$$

(the kinetic energy $T_p^{\text{thr.}} = 5.64$ GeV). However, the experiments [*30, 31*] detect antiprotons production acts up to the energies $T \approx 3$ GeV.

It is clear that the sub-threshold heavy pair production process is an extreme situation which also requires a special consideration since for its description it makes evidently no sense to use scale properties of the inclusive spectrum as energies are too small.

Nevertheless, for its interpretation one still can use the concepts of the "gathering" model. However, one should apply its modification suitable for the relevant analysis. According to this modification the relative yield of antiprotons may be represented by the approximate formula [*9*]:

$$\varphi = A \cdot W^{(n)} \cdot \psi(p_{\bar{p}}) \cdot \eta(p_{\bar{p}}). \tag{75}$$

Three factors in (75), $W^{(n)}$, $\psi(p_{\bar{p}})$ and $\eta(p_{\bar{p}})$ correspond to the three phases of the process. The factor $W^{(n)}$ is related to the first phase, the "gathering" of n nuclear nucleons into a compound-system and, as before, has the meaning of the process cross-section. The $W^{(n)}$ are computed for the parameter set (59). $\psi(p_{\bar{p}})$ is the probability of the system decay

(i.e. the second phase of the process) accompanied by the production of an antiproton with momentum $p_{\bar{p}}$. For the description of $\psi(p_{\bar{p}})$ in [9] the thermodynamical model was used. The approximate estimation which takes into account.
a) the fact of observation of an antiproton at $\theta \approx 0°$ and
b) a possibility of the choice of energy $E_p^{(n)}$ so that the antiproton with momentum $p_{\bar{p}}^0 \approx \langle p_{\bar{p}}^0 \rangle \approx 0.45$ GeV/c in the center-in-rest system, have momentum $p_{\bar{p}}$ in laboratory reference frame near the value fixed experimentally, gives the simple expression for function

$$\psi(p_{\bar{p}}) \approx \frac{p^2}{\sqrt{p^2 + m^2}} \sqrt{p_0^2 + m^2} \exp\left(-\frac{p_0^2}{2mT}\right), \tag{76}$$

where $T = \mu_\pi$ is the temperature and

(77)

$$p_{\bar{p}} \approx (p_{\bar{p}}^0 + V_{(n)} E_{\bar{p}}^0) \cdot \gamma_{(n)}. \tag{78}$$

As before, $\gamma_{(n)}$ and $V_{(n)}$ are determined by kinematics of the "gathering" mechanism:

$$V_{(n)} = \frac{\sqrt{E_p^{(n)2} - m^2}}{E_p^{(n)} + m \cdot n}; \qquad \gamma_{(n)} = \frac{E_p^{(n)} + nm}{\sqrt{m^2(n^2 + 1) + 2nmE_p^{(n)}}}. \tag{79}$$

The energy $E_p^{(n)}$ is taken from relation [9]:

$$E_p^{(n)} \cong \frac{1}{n} [m(3n + 4)(n + 3)\,\overline{T}_{p,\bar{p}}] \tag{80}$$

($\overline{T}_{p,\bar{p}} \approx 2\overline{T}_p \approx 0.2$ GeV is the total average kinetic energy of the pair in the cluster-in-rest system).
This method allows us to avoid difficulties appearing in the consideration of near-threshold phenomena.
Finally a few words about $\eta(p_{\bar{p}})$ which represents the latter phase of the process: passing of antiprotons through a nucleus. We take into account, roughly, a possibility of the antiproton annihilation. Note, first of all, that the cumulative process for the chosen parameters σ and $\bar{\tau}_0$ possesses the property to be proportional to nuclear volume (see III, 1). So the "survival" of antiprotons in matter in first approximation is to be proportional to the width of nuclear layer transparent with respect to annihilation at its back surface.
The transparent layer width approximately equals the mean free path of an antiproton with respect to annihilation. So we assume that the "surviving" factor is proportional to the expression:

$$\eta(p_{\bar{p}}) \sim 1/\varrho\sigma_{\text{ann.}}(p_{\bar{p}}). \tag{81}$$

Data on the cross-section of antiproton annihilation in $\bar{p}$, N-interaction [32] may be represented by the relation:

$$\sigma_{\text{ann.}}(p_{\bar{p}}) \approx (63 \pm 2) \cdot p^{-(0.64 \pm 0.02)} \text{ mb}. \tag{82}$$

For estimating φ für $n \geqq 2$, in accordance with (80), $E_p^{(n)}(T_p^{(n)})$ and then $V_{(n)}$, $\gamma_{(n)}$ and $W^{(n)}$ are calculated. Choosing momenta $p_{\bar{p}}$ the same as in experiment [31] at energies T_p close to $T_p^{(n)}$ one can calculate the functions $W^{(n)}$ and $\eta(p_{\bar{p}})$.

The results are represented in fig. 20 together with the expressental data. Normalization $\varphi_{\text{theor.}}$ has been produced via the choice of the point at $T_p = 6$ GeV and $p_{\bar{p}} = 2.0$ GeV/c, so that $\varphi_{\text{theor.}} = \varphi_{\text{exp}}$ in the above-threshold range ($T_p > 5.64$ GeV).
Let us stress once more that this normalization is justified, as it is based on the correct description of qualitative composition of particles in the thermodynamical model.
Fig. 16 demonstrates the satisfactory agreement between the model and experiment: the rate of φ falls off with the T_p decrease, the arrangement and spread of points corresponding to various momenta at $T_p =$ const. do not contradict the observations.

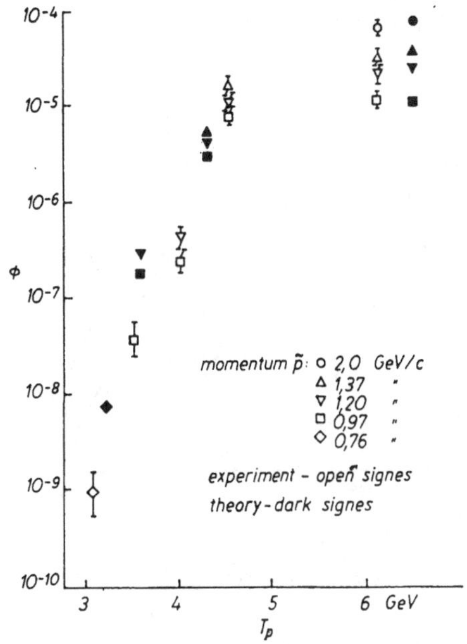

Fig. 16. Ratio $\varphi(T_p)$ for the nucleus-target Cu

6. On "cumulative" protons

We have not compared the model with the experiment on protons flying into the backward hemisphere. It seems that such a comparison is premature, as in contrast to the processes leading to production of new particles, protons available in a nucleus in a "prepared" form may leave the nucleus due to the action of nonrelativistic, purely nuclear mechanisms. A definite contribution may come in this case from rescatterings of nonrelativistic recoil nucleons accompanying the high energy act of multiple production and also by other mechanisms. In this connection the results of work [*33*] are of interest.
Being the sum of contributions both of the relativistic mechanism of hadron collective interaction and of nonrelativistic channels caused by the inevitable appearence of a large number of secondary nucleons thrown off by an excited nucleus, the process $\text{h} + A \to \text{p}(\theta = 180°) + X$ turns out to be insufficiently "clear" for studying a truly cumulative effect. Here additional investigations beyond the scope of the cumulative scheme study are necessary.

V. Some "predictions" for future experiments

To establish (in the framework of adopted approximations) the range of applicability and possible refinement of the formulated picture of cumulative processes, it is useful to extend the studies by including the data on production of a number of other cumulative particles and also on the dependence of their yield on energy in a wide interval.
For this purpose, the calculations — predictions were carried out for invariant production cross-sections in the processes which have not yet been studied experimentally. In fig. 17 $R_A(\theta = 180°, T)$ is given for process $\mathrm{p} + A \to (\Lambda°, \Sigma) + X$; $P^{\mathrm{in}} = 8.6$ GeV/c. In

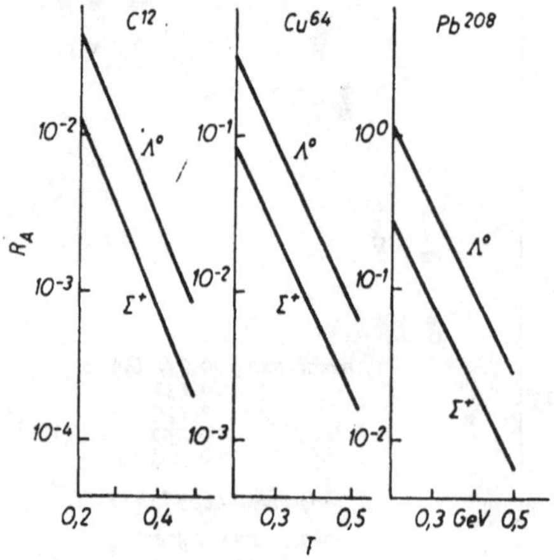

Fig. 17. $R_A(\theta = 180,° T)$ for the process $\mathrm{p} + A \to (\Lambda°, \Sigma) + X$; $P^{\mathrm{in}} = 8{,}6$ GeV/c

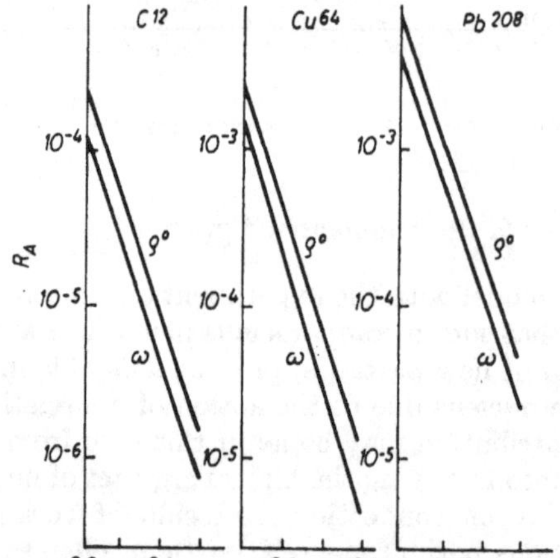

Fig. 18. $R_A(\theta = 180,° T)$ for the process $\mathrm{p} + A \to (\mathrm{p}°, \omega) + X$; $P^{\mathrm{in}} = 8.6$ GeV/c

fig. 18 R_A obtained, however, at rather large simplifications (see [*10*]) are represented for $\rho°$ and ω-mesons. This is due to the lack of the necessary information on the invariant density behaviour in pN-process. In fig. 19 R_A are depicted for $p + A \to (\mathrm{K}^{\pm}, \bar{\mathrm{p}}, \theta = 180°) + X$, $P^{\mathrm{in}} = 8.6$ GeV/c. And finally, fig. 20 demonstrates the results for $\mathrm{p} + A \to (\pi^-, \mathrm{K}^-, \bar{\mathrm{p}}; \theta = 160°) + X$, $P^{\mathrm{in}} = 400$ GeV/c for the parameters set (61) that, as has been said above, seems preferable to us. In addition let us make one more remark. The condition

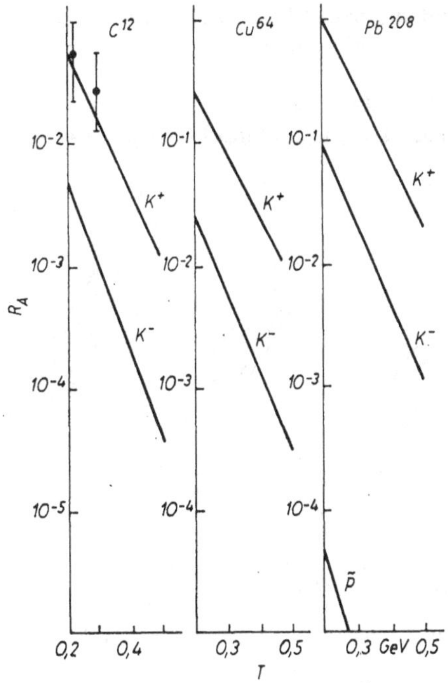

Fig. 19. $R_A(\theta = 180°, T)$ for the process $p + A \to (K\pm, \bar{p}) + X$; $P^{in} = 8.6$ GeV/c

(62) implies that the system before its decay approaches the state in which the energy is distributed over the hadron volume and whose decay may be described thermodynamically.

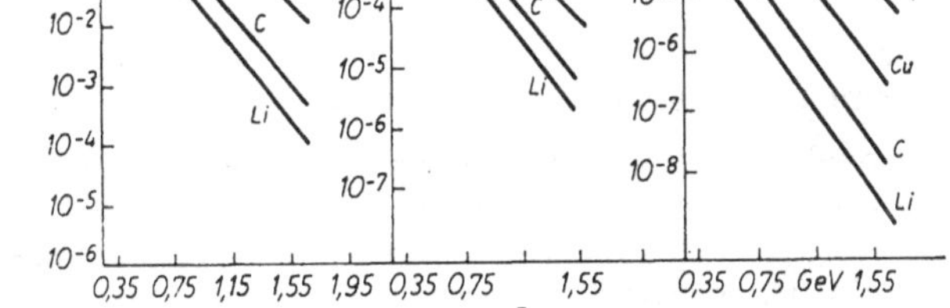

Fig. 20. $R_A(\theta = 160°, T)$ for the process $p + A \to (\pi^-, x^-, \bar{p}) + X$; $P^{in} = 400$ GeV/c, parameter set (61)

The same evaluation for the parameter set (59) will produce for $\langle r \rangle$ the value three times smaller as compared to (62). Hence, in this case we would deal with the situation appropriate to gluon and quark hard scattering. (Then perhaps we should reformulate our model in its kinematics and in definition of cross sections since in this case we should describe not the collision of hadron as unique objects but of their constituents).

However, it is known that picture of hard scatterings does not yet provide the interpretation of the inclusive cross section of large $p_\perp$-particle production ($p_\perp < 6-7$ GeV/c) at $P^{\text{in}} = 400$ GeV/c [*34*].
The above mentioned arguments show that in this very preliminary study of the problem the parameters set (61) seems to be preferable for process at $P^{\text{in}} = 400$ GeV/c.

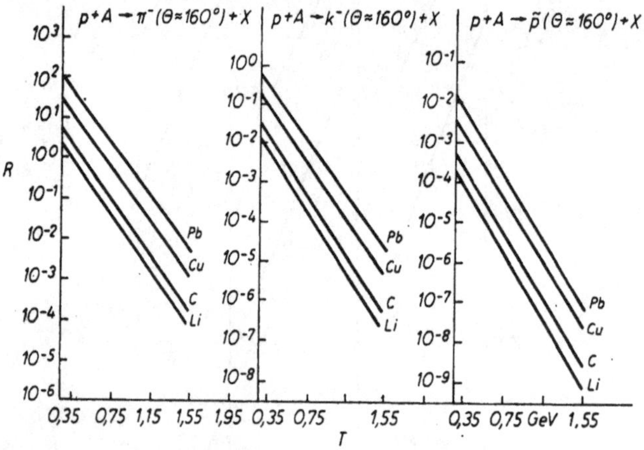

Fig. 21. $R_A(\theta = 160°, T)$ for the process p + $A \to (\pi^-, x^-, \tilde{p}) + X$; $p^{\text{in}} = 400$ GeV/c, parameter set (59)

It stands to reason the decisive factor is the future experiment.
To complete this discussion note that we have obtained also the spectra of π^--, K^--mesons and $\tilde{p}$-antiprotons in both the variants and at other angles in the backward hemisphere (up to $\theta = 90°$). Here we do not cite these results because it will increase considerably the volume of the paper.
Comparing fig. 19 and fig. 20, one can reveal the sharp increase of $R_A^{(\tilde{p})}$ when passing to higher energies. According to III, § 3, this effect should indeed take place in the production of a heavy mass or a heavy pair. This, it should be expected that the observation of cumulative antiprotons being extremely difficult at $P^{\text{in}} = 8 \div 10$ GeV/c, may become possible at $P^{\text{in}} = 400$ GeV/c.

VI. Conclusion

Now we shall formulate briefly a number of statements which can be considered as a result of represented cumulative-effect investigations.

1. The "gathering" model is phenomenological, as it does not use directly the details of hadron structure and also does not need any details of dynamics of the development and decay of a compound-system. Only general principles of identity or of scale invariance of inclusive spectra and also the parameters describing the compound-system-production cross section and average time of produced-particles formation are used in it.
Certainly, it is very tempting to assert that the process is determined by the objects just as quarks and gluons.
However, the manifestation of cumulative effects in various reactions is complicated enough, and the role of many factors is not yet cleared. On the other hand, the quark-glue concept still requires a significant refinement.
So, in our opinion, at the given stage the phenomenology is not the drawback of the given model.

Establishing basic physical quantities with its help, determining the process gives a possibility to realize just what is to be explained further, by using detailed ideas on the hadron structure.

In other words, the phenomenological model helps to know what results do not depend on structure details and to extract the relationships or factors which are to become subjects of investigation of the structure theory. As is known, a good phenomenology simplifies significantly the formulation of theory based on a more deep penetration into the structure.

From this point of view, one may state that the phenomenology formulated in the framework of the "gathering" mechanism describing well the most general features of cumulative effects is by itself a useful result. Besides, in the range $P^{\text{in}} \lesssim 40$ GeV/c the picture appears to be universal.

2. The model is formulated on the basis of general consequences arising from the necessary consideration of space-time regions of the process developing.

It should be emphasized once more that the model discussed does not need the presence in a nucleus of a ready object of double, triple, and so on nucleon mass, i.e. does not require additional assumptions on the nucleus structure. It proceeds from the inevitability of appearing of such an object by using the apparent fact: any physical process, for its realization, requires finiteness of the space-time interval. The latter, for dense nuclear medium, when not one nucleon can be placed in the gathering range, leads to appearing of intermediate states with masses much greater than the maximal probable mass in pN-interaction at the same incident energy. In other words, it should be surprised if such a mechanism does not contribute to cumulative processes.

So, as this mechanism not only explains basic features but also reproduces correctly the magnitude of the cumulative particles yield, it may pretend to play a role of the dominant one.

3. Moreover, the analysis of available experiments by means up to model has allowed us to extract the information important for structure models.

a) The time estimation of formation of cumulative particles in a compound-system has been derived. From this estimation it follows that longitudinal sizes of the range of interaction leading to the cumulative production do not increase with growing energy (in c.m.s.).

b) In a wide energy interval 2 GeV $< E <$ 400 GeV the production cross-section has been defined for the compound-systems capable to concentrate a significant part of its energy for emission of a single particle.

This information has been obtained for the first time. It seems, any picture of hadron structure is not capable to explain such a behaviour of a compound-system-production cross-section.

c) Available experimental data on the cumulative range do not contradict the identity hypothesis (or for large energies, scale invariance) of inclusive spectrum in hadron-hadron and collective-hadron interactions.

We believe that here there also are possible interesting consequences. For instance, as we have already seen, in the collective interaction an intermediate system covers a more complicated path before the decay (acts of mass enlarging are distributed in time). However, when masses are equal, the results of decay for both the cases are practically the same.

In particular, it may imply that not only the initial (parton) spectrum of colliding particles (as is commonly believed) plays an important role in formation of produced hadrons spectrum, but also the character of hadronization process, i.e. the phase transition of the quark-glue matter into hadrons, by itself, occuring at the final stage of the reaction.

So, the study of cum ılative processes in the direction described leads to quite interesting and important, for nuclear physics, results. Hence further investigation is extremely desirable.

One may point out the following rather interesting and capacious problems: the direct use of conclusions for improving the hadron-structure models, further development of the cumulation model itself, e.g., its extension to the case of relativistic nuclei interaction.

In conclusion, the authors express their gratitude to A. M. Baldin for the talks helping to formulate the model, to V. S. Stavinsky for explanations of the experimental situation, and also to V. I. Ogievetsky, A. V. Efremov, B. N. Valuev, M. I. Shirokov, I. V. Polubarinov and L. G. Zastavenko — for very useful discussions of a number of problems and results and for constructive comments.

References

[1] A. M. BALDIN, JINR E2-9138, Dubna (1975); Proc. of Conf. on High Energy Physics and Nuclear Structure-Santa Fe, 373, 1975; A. M. BALDIN, Elementary Particles and Atomic Nuclei, 8, 429 (1977).
[2] G. A. LEKSIN, Nuclear Scaling, MIFI, Atomizdat (1975).
[3] J. W. CRONIN, Phys. Rev. **D 11**, 3105 (1975).
[4] D. ANTREASYAN et al., Phys. Rev. Letters **38**, 112; 115 (1977); L. KLUBERG et al., Phys. Rev. Letters **38**, 670 (1977).
[5] T. HAYASHINO et al., Lettere Nuovo Cimento **16**, 71 (1976).
[6] N. ANGHELOV et al., JINR P1-11951, Dubna (1978).
[7] B. N. KALINKIN, A. V. CHERBU, V. L. SHMONIN, JINR, P2-10783, Dubna (1977); Acta Phys. Polon. **B 9**, 375 (1978).
[8] B. N. KILINKIN, A. V. CHERBU, V. L. SHMONIN, JINR P2-10784, Dubna (1977); Acta Phys. Polon. **B 9**, 385 (1978).
[9] B. N. KALINKIN, A. V. CHERBU, V. L. SHMONIN, JINR P2-10785, Dubna (1977); Acta Phys. Polon. **B 9**, 393 (1978).
[10] B. N. KALINKIN, A. V. CHERBU V, L. SHMONIN, JINR P2-11621, Dubna (1978).
[11] B. N. KALINKIN, A. V. CHERBU, V. L. SHMONIN, JINR P2-11622, Dubna (1978).
[12] B. N. KALINKIN, A. V. CHERBU, V. L. SHMONIN, JINR P2-11802, Dubna (1978).
[13] B. N. KALINKIN, V. L. SHMONIN, JINR P2-11380, Dubna (1978); Z. Phys. **A 238**, 227 (1978).
[14] B. N. KALINKIN, S. N. KOLTOCHNICK, V. L. SHMONIN, HEPI 61-78, Alma-Ata (1978); D. G. BARANOV et al., FTI Ioffe, 562, Leningrad (1977); D. G. BARANOV et al., Pizma Zh. eksper. teor. Fiz., 28, 475 (1978).
[15] B. N. KALINKIN, A V. CHERBU, V. L. SHMONIN, JINR P 2-11620, Dubna (1978).
[16] E. L. FEINBERG, Uspechi fiz. Nauk **104**, 539 (1971).
[17] A. M. BALDIN, JINR P 7-5769, Dubna (1971); A. M. BALDIN, Comm. Phys. No. 1, USSR AS, 35 (1971).
[18] B. M. ABRAMOV et al., Pizma Zh. eksper. teor. Fiz. **26**, 403 (1977).
[19] A. M. BALDIN et al., Zh. eksper. teor. Fiz. **18**, 79 (1973).
[20] I. A. RADKEVICH, Elementary Particles, III, 104, Atomizdat, ITEP (1973).
[21] V. S. BARASHENKOV, et al., Yadern Fiz. **13**, 743 (1971).
[22] N. N. NIKOLAEV, V. N. ZAKHAROV, Yadern Fiz. **21**, 434 (1975).
[23] J. H. CHRISTENSEN et al., Phys. Rev. **D 8** 2016 (1973).
[24] A. M. BALDIN, et al., Contribution to the European Conf. on Particle Physics, Budapest (1977).
[25] A. M. BALDIN et al., JINR 1-8249, Dubna (1974).
[26] G. A. LEKSIN, A. V. SMIRNITSKY, Pizma Zh. eksper. teor. Fiz. **28**, 179, (1978).

[27] G. A. Leksin, A. V. Smirnitsky, Pizma Zh. eksper. teor. Fiz. **28**, 97 (1978).
[28] G. Fox, XIX Int. Conf. High Energy Physics, Tokyo (1978).
[29] C. Michael, C. Wilkin, Nuclear Phys. **B 11**, 99 (1969).
[30] T. Elioff, Phys. Rev. **128**, 869 (1962).
[31] D. E. Dorfman et al., Voprosi fiziki elementarnich chastits, Erevan, 1966.
[32] V. Simak, Internat. Seminar on Deep Inelastic and Multiple Production Processes by High Energies, Dubna, 1973; G. Giacomelli, 16-th Internat. Conf. on High Energy Physics, Chicago and Batavia, 1972.
[33] K. K. Gudima, et al., JINR E2-11307 (1978).
[34] B. L. Combridge et al., Phys. Letters **70 B**, 234 (1977).

INHALT

Die „FORTSCHRITTE DER PHYSIK" dienen der Publikation zusammenfassender Berichte über aktuelle Teilgebiete der Physik. Neben Originalberichten in deutscher oder englischer Sprache werden deutsche Übersetzungen wichtiger fremdsprachiger Berichte abgedruckt.

Aus dem Inhalt der folgenden Hefte:

H. MATSUMOTO, G. SEMENOFF, M. TACHIKI, H. UMEZAWA: Quantum Electrodynamics in Solids

W. RÜHL, D. LAZO-TUEVE: Invariant Forms of the Lorentz Group

A. A. GRIB, S. G. MAMAYEV, V. M. MOSTEPANENKO: Vacuum Stress-Energy Tensor and Particle Creation in Isotropic Cosmological Models

F. SCHWARZ: An Approcimation Scheme for Constructing $\pi^0\pi^0$ Amplitudes from ACU Requirements

K. J. BIEBL, M. KLEIN, R. NAHNHAUER: A Jet Model Study of Correlations in Hadron-Hadron Reactions Due to Resonance Production

J. FISCHER, R. ŠÁLY: The Pomeranchuk Theorem and Its Modifications

J. LUKIERSKI: Field Operator for Unstable Particle and Complex Mass Description in Local QFT

P. MOYLAN: Fiber Bundles in Non-Relativistic Quantum Mechanics

Manuskripte werden nach Vereinbarung mit dem Autor von der Schriftleitung entgegengenommen.

FSC
www.fsc.org
MIX
Papier | Fördert gute Waldnutzung
FSC® C083411

Zeitfracht Medien GmbH
Ferdinand-Jühlke-Straße 7
99095 Erfurt, Deutschland
produktsicherheit@kolibri360.de